Engineering Materials

This series provides topical information on innovative, structural and functional materials and composites with applications in optical, electrical, mechanical, civil, aeronautical, medical, bio- and nano-engineering. The individual volumes are complete, comprehensive monographs covering the structure, properties, manufacturing process and applications of these materials. This multidisciplinary series is devoted to professionals, students and all those interested in the latest developments in the Materials Science field.

More information about this series at http://www.springer.com/series/4288

Rafael Chaves Lima · Lindiane Bieseki ·
Paloma Vinaches Melguizo ·
Sibele Berenice Castellã Pergher

Environmentally Friendly Zeolites

Synthesis and Source Materials

 Springer

Rafael Chaves Lima
Federal University of Rio Grande
do Norte – UFRN
Natal, Brazil

Lindiane Bieseki
Federal University of Rio Grande
do Norte – UFRN
Natal, Brazil

Paloma Vinaches Melguizo
Brazilian Synchrotron Light Laboratory
Brazilian Center for Research
in Energy and Materials
Campinas, São Paulo, Brazil

Sibele Berenice Castellã Pergher
Federal University of Rio Grande
do Norte – UFRN
Natal, Brazil

ISSN 1612-1317   ISSN 1868-1212 (electronic)
Engineering Materials
ISBN 978-3-030-19972-2   ISBN 978-3-030-19970-8 (eBook)
https://doi.org/10.1007/978-3-030-19970-8

This Springer imprint is published by the registered company Springer Nature Switzerland AG
The registered company address is: Gewerbestrasse 11, 6330 Cham, Switzerland

Foreword

The synthesis of zeolites is a mature branch within the field of materials. However, still the synthesis of zeolites is an area of knowledge in constant evolution and development. There is no doubt that the main driver of this constant interest in the synthesis of zeolites comes from chemistry and petrochemicals since zeolites are the most used solid catalysts in the refining industry. Indeed, the control of the properties of the zeolites has allowed to greatly improving the selectivity of many chemical processes and therefore to diminish their environmental impact and to increase productivities of the units in which the zeolites are applied as catalysts. No less important than its practical application is the academic interest that this type of material arouses in catalysis and adsorption. This is because this type of materials allows very precise control of those properties whose effect is intended to be studied in catalytic processes of scientific interest. Thus, it is possible to control the strength and number of Bronsted acid centres by incorporating trivalent elements into the zeolitic framework. Likewise, the incorporation of other heteroelements, such as Ti, Sn, Zr, provides redox properties or Lewis acidity to the zeolitic catalysts. The right selection of the amount and type of heteroatom incorporated to the zeolitic catalyst allows controlling very precisely the type of reaction that is envisioned to catalyse.

If the presence of heteroatoms other than silicon in the zeolitic catalysts is of paramount importance, the topology of the zeolite is also fundamental. Thus, the pore size that gives access of reagents to the inner surface area of zeolites permits to control which molecules can diffuse and reach the active catalytic centres of zeolites, and which will not have access, behaving as true molecular sieves. Until now, 245 zeolitic topologies have been described with a wide variety of channels and cavities, providing them of a large suite of catalytic and adsorption properties.

Thus, it can be concluded that the use of zeolites in catalysis and in adsorption processes is a field with a long scientific and technological path ahead, despite the time elapsed since its discovery and first industrial uses.

Therefore, a book like the one presented here has a double objective. On the one hand, it is an excellent introduction to the field of zeolites for students and researchers who start in this area. On the other hand, it is an excellent compendium about the parameters that govern the zeolite crystallization. This information is very often scattered and difficult to find.

Thus, Chaps. 1 and 2 are remarkable, in which a historical perspective of the zeolites is provided, as well as a detailed structural description of the same is given. Finishing, this second chapter with an excellent section dedicated to natural zeolites. This volume continues, describing exhaustively the parameters generally taking into account during the zeolite preparation, such as gel composition, temperature and time of crystallization, mobilizing agent, stirring. While Chap. 4 describes synthetic approaches that reduce the environmental impact of the production of zeolites. This is mostly achieved by optimizing the synthesis methods using industrial, mining or agricultural wastes as raw materials for the zeolite crystallizations. In this way, industrial wastes are valourized, while the environmental footprint in the production of zeolites is reduced.

All these aspects allow reaching a high understanding of the state of the art in the synthesis of zeolites and permit to reach a high rationalization for preparation of zeolites to the scientists who develop their research in this field. I hope that this topical volume is as useful for them as it has been for me.

Valencia, Spain   Dr. Fernando Rey
                                                    Director of Instituto de Tecnologia
                                                                Química (ITQ)

Preface

Zeolite synthesis is a very interesting research area, and these important materials, obtained in 245 different structures, can be employed in a variety of processes. In this book, we describe what zeolite is, the structures of zeolites and the parameters that influence zeolite synthesis, thereby showing a new perspective of this field. Next, different processes are shown that can be employed to synthesize zeolites using residues, natural materials and other eco-friendly materials, such as raw powder glass, clays, aluminium cans, diatomites, rice ashes and coal ashes. Finally, the objective of this book is to give the reader a wide range of possibilities for synthesizing zeolites in eco-friendly ways, so that these options can be applied to different industrial processes.

We hope that readers enjoy this book and that it will be useful for future research.

Natal, Brazil   Rafael Chaves Lima
Natal, Brazil   Lindiane Bieseki
Campinas, Brazil   Paloma Vinaches Melguizo
Natal, Brazil   Sibele Berenice Castellã Pergher

Introduction

Zeolite materials are solid porous materials known as molecular sieves because of their capacity to select molecules by size. These materials can be applied to several processes, such as catalysis, adsorption and separation.

The importance of zeolites lies in the different structures that can be synthesized with different compositions. The website of the International Zeolite Association (http://www.iza-online.org/) provides a large amount of information about the structures of these materials, how they are synthesized, references and XRD patterns. Currently, 245 structure types have been recognized; each structure has a code of three letters, for example, faujasite zeolite has the code FAU. Extensive research has been conducted searching for new structures and known structures with new compositions. However, only 11 zeolites have been produced and applied industrially because of the cost of synthesis. Thus, the study of more economic zeolite synthesis routes is very important. Studies of scaled up synthesis, the optimization of the synthesis parameters and the use of natural materials and wastes as sources of Si and Al are fundamental for the industrial production of other zeolites.

To achieve eco-friendly zeolite synthesis, it is important to understand and control zeolite synthesis. This book describes what zeolite is, how these materials can be synthesized, how the parameters influence zeolite formation and eco-friendly synthesis.

Contents

1   **Zeolites: What Are They?**   1
    1.1   Zeolite Memories   1
    1.2   Concept   5
    1.3   Zeolite LEGO: Structure and Taxonomy of Zeolites   6
    1.4   Natural and Synthetic Zeolites   13
    References   15

2   **Zeolite Synthesis: General Aspects**   21
    2.1   Synthesis Methods   21
    2.2   Physical Chemistry of Zeolite Crystallization   22
    2.3   Mechanisms of Zeolite Formation   29
    2.4   Influence of Synthesis Parameters   42
        2.4.1   Temperature   43
        2.4.2   Gel Aging   44
        2.4.3   Pressure   45
        2.4.4   Stirring or Static Thermal Treatment   45
        2.4.5   Chemical Composition of the System   45
        2.4.6   Basicity and Acidity of the Synthesis Medium   47
        2.4.7   Sources of Components   48
        2.4.8   Time   56
    References   58

3   **Zeolite Eco-friendly Synthesis**   65
    3.1   Coal Ash Raw Materials   65
    3.2   Rusk Rice Ash Raw Materials   70
    3.3   Clay Raw Materials   71
    3.4   Raw Powder Glass Materials   74
    3.5   Industrial Wastes as Raw Materials   78
    3.6   Other Raw Materials   80
    References   85

4 Recipes of Some Ecofriendly Syntheses 93
   4.1   Zeolites from Coal Ash Raw Material 94
   4.2   Zeolites from Rusk Rice Ash Raw Material 96
   4.3   Zeolites from Clay Raw Material 96
   4.4   Zeolites from Raw Powder Glass Materials 100
   4.5   Zeolites from Other Raw Materials 106
   References 109

Conclusions: Future Approaches 111

About the Authors

Rafael Chaves Lima obtained his Master degree in Chemistry from the Federal University of Rio Grande do Norte (UFRN-Brazil). Nowadays, he is pursuing a doctoral degree in Chemistry under the supervision of Dr. Sibele Pergher in the same university. His research is focused on Inorganic Physical Chemistry, specifically on zeolite synthesis.

Dr. Lindiane Bieseki obtained in 2016 her Ph.D. degree in Materials Science and Engineering from the Federal University of Rio Grande do Norte (UFRN, Brazil) under supervision of Dr. Sibele Pergher. Currently, she is researching about zeolites at LABPEMOL (UFRN). She acts mainly on the following topics: acid treatments and pillarization of clays and zeolite synthesis, with emphasis on their obtention from natural raw materials and residues containing silica and alumina in their compositions.

Dr. Paloma Vinaches Melguizo does research in Materials Chemistry, Catalysis and Inorganic Chemistry, focused on zeolite synthesis and characterization. She defended her doctoral thesis on zeolites structure-direction in 2017 under the supervision of Dr. Sibele Pergher at the Federal University of Rio Grande do Norte (UFRN). Nowadays, she holds a post-doctoral position in the group of Dr. Florian Meneau at Brazilian Synchrotron National Laboratory (LNLS/CNPEM) studying zeolites with synchrotron techniques.

Dr. Sibele Berenice Castellã Pergher is Chemical Engineering (UFRGS-1990), Masters in Chemical Engineering (UEM-1993), Doctor in Chemistry (UPV/ITQ-Spain–1997). She worked at UFRGS (1998–2001) and at URI—Erechim (2001–2010). She is currently a professor and researcher (since 2010) at UFRN. She is also director of the Brazilian Society of Catalysis—SBCat, part of the Synthesis Commission of IZA, represents Brazil in FISOCAT and IACS. She is coordinator and founder of LABPEMOL. And she works mainly on synthesis of catalysts, zeolites, clays, mesoporous materials, lamellar, pillared and delaminated materials, adsorption and catalysis processes. She has more than 150 published papers, 20 patents and 500 work on Congress. She has a great contribution to academic student formation, with more than 150 orientation concluded.

Chapter 1
Zeolites: What Are They?

In this chapter, the main aspects of the study of zeolites including the chronology of the research, structural descriptions and classification of the zeolitic materials are described.

1.1 Zeolite Memories

The word zeolite, from the Greek words ζειν, "zeo" or "zein" (boiling) and λιθοσ or "lithos" (stone), i.e., "boiling stone", was adopted in 1756 by the Swedish mineralogist Baron Axel Fredrik Crönstedt in his publication *An Essay Towards a System of Mineralogy* (Masters and Maschmeyer 2011). The First zeolites were minerals whose composition consist of hydrated aluminosilicates, first studied by the same researcher when stilbite was discovered, although there is no proof that the material studied was effectively this zeolite. Thus, it is thought that chabazite was the first zeolite discovered, as described in 1792 by Louis Augustin Guillaume D 'Antic (Colella 2005; Millini and Belussi 2017).

The use of this term stemmed from the perception that when this material presented intumescence, that is, when this material was heated with borax, it simultaneously released vapors (Corrêa et al. 1996; Smith 2000; Guisnet and Ribeiro 2004).

In 1840, A. Damour studied the reversible dehydration capacity of natural zeolites without any apparent structural damage (Flanigen 2001; Colella 2005). In 1858, the German chemist H. Eichhorn reported reversible ion exchanges in natrolite and chabazite zeolites (Millini and Belussi 2017). In 1862, Henry Etienne Sainte-Claire Deville reported the first synthesis of a zeolite, levynite, by heating an aqueous solution of potassium silicate and sodium aluminate in a glass tube at 170 °C (Gottardi and Galli 1985; Masters and Maschmeyer 2011). Finally, in 1896, Georges Friedel suggested that the structure of a dehydrated zeolite resembled a sponge by observing that certain molecules could be occluded in this material (Flanigen 2001; Lazlo 2018).

© Springer Nature Switzerland AG 2019
R. Chaves Lima et al., *Environmentally Friendly Zeolites*, Engineering Materials,
https://doi.org/10.1007/978-3-030-19970-8_1

In the 19th century, great contributions to the science of these materials were presented. Robert Gans used zeolites for water removal, giving commercial importance to these materials (Masters and Maschmeyer 2011). In 1909, François Grandjean observed the adsorption of chemical species in dehydrated chabazite (Flanigen 2001). In 1924, Oskar Weigel and Edward Steinhoff first described molecular selectivity in zeolites by observing that chabazite adsorbed small chain alcohols and formic acid but not propanone, ether or benzene (Weigel and Steinhoff 1924). W. H. Taylor and Linus Pauling determined the first zeolitic framework structures in 1930 (Masters and Maschmeyer 2011; Zimmermann and Haranczyk 2016), although Breck (1964a) said that the first structural analysis was reported by himself and T. Reed in 1956.

In 1932, James W. McBain introduced the term molecular sieve to describe porous materials that select the incorporation of molecules in the material by molecular size (Flanigen 2001). From then until the 1940s, Richard Maling Barrer conducted a systematic and pioneering study regarding various aspects of zeolites, including the adsorption of gases and the conversion of known minerals using strong salt solutions at high temperatures. The zeolitic "synthetic era" began (Breck 1964b; Flanigen 2001; Cundy and Cox 2003).

In 1948, the first synthetic zeolite, synthetic mordenite, was reported, which was analogous to natural mordenite (Barrer 1948). In the period from 1949 to 1954, approximately twenty structures, including aluminum-rich zeolites A (LTA) and X (FAU), were obtained by a research group composed of well-known scientists such as Robert M. Milton, Thomas B. Reed and Donald W. Breck from the current Union Carbide Corporation (Breck et al. 1956; Flanigen 2001).

In the 1950s and 1960s, the zeolitic "industrial era" began (Byrappa and Yoshimura 2013). Research on the properties of these materials increased, and the potential uses of zeolites in industrial processes became more evident given the large number of patents related to synthesis and application (Fig. 1.1) that arose during that time.

In this period, several zeolites with higher silicon contents were synthesized and marketed, such as zeolite Y (FAU) prepared by the Union Carbide Corporation (Breck 1964b) and high porosity mordenite marketed as Zeolon by the Norton Company (Szostak 1992).

From the 1950s, the Mobil Oil Corporation began to introduce zeolites in certain newer applications and to find new zeolitic structures. In 1958, this company successfully tested organic cations in zeolite synthesis, obtaining new structures. The effect of quaternary ammonium cations on zeolite synthesis was also studied in 1961 by Mobil researchers (Cundy and Cox 2005). In 1962, the same company introduced zeolite X containing alkali earth metal cations as a component of a cracking catalyst.

In the 1960s, the synthesis of silicon-rich zeolites such as zeolite ZK-4, a version of LTA with higher silicon content, was recorded by Kerr (1966). Zeolite beta (Wadlinger et al. 1967) and ZSM-5 (Argauer and Landolt 1972) were also obtained during this period using organic cations as structure-drivers. At this time, chemical modification of the zeolite Y to ultrastable form (USY) was also reported by the Grace Company (McDaniel and Maher 1969; Flanigen et al. 2010). In 1967, the first international conference of zeolites (IZA) took place (Flanigen 2001).

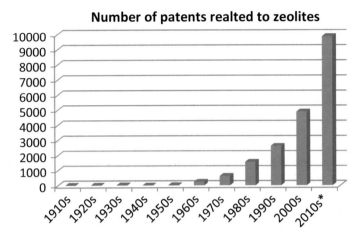

Fig. 1.1 Number of patents related to zeolites by decade according to Scifinder® database

In 1974, the German company Henkel introduced zeolite A in detergents as a substitute for the environmentally harmful polyphosphates used at the time for use with hard water (Yamane and Nakazawa 1986; Flanigen et al. 2010). In 1978, Edith Marie Flanigen and coworkers introduced a new zeolite composition called pure silica, such as the silicalite family, by synthesis of silicalite using fluoride ions (Flanigen and Patton 1978).

In the 1980s, forty natural zeolites were identified, and more than 10,000 patents related to zeolite synthesis were presented (Giannetto et al. 2000). In 1982, Union Carbide research groups reported a large increase in the number of known zeolitic structures through the synthesis of phosphate-based zeolites, called zeotypes. A group led by Stephen T. Wilson reported on aluminum phosphate (AlPO) synthesis (Wilson et al. 1983). One year later, the group of Brent M. Lok reported the synthesis of silicoaluminophosphates (SAPOs) (Lok et al. 1984). Furthermore, the substitution of Ti atoms in silicalite-1, generating TS-1 titanosilicate, was reported by M. Taramasso and collaborators (Taramasso et al. 1983).

The 1980s also yielded research aimed at postsynthesis chemical modifications and mathematical models applied to microporous materials (Flanigen 2001; Cundy and Cox 2005).

Subsequently, many new zeolite structures have been reported, for example, cloverite, a galophosphate obtained by Estermann et al. (1991) and a series of zeolites named ITQ-X (X = number) produced by Avelino Corma and several collaborators since the late 1990s (Camblor et al. 1998; Corma et al. 1998; Bieseki et al. 2018).

Recent years of zeolite development have been characterized by the addition of new processes and fields of application and the study of controlling zeolite properties and morphologies, which has so far culminated in nanosized zeolites (Mintova et al. 2016), hierarchical zeolite molecular sieves (Möller and Bein 2011) and novel routes to obtain new zeolite and zeolite-like structures (Morris and Čejka 2015).

Although natural zeolites were the first type of these interesting materials to be studied, the large and still growing group of synthetic zeolites has predominated in terms of research. Approximately 60 species divided into 28 framework types were discovered in nature (Stöcker et al. 2017) of the approximately 230 known zeolite topologies (Baerlocher and McCusker 2018), all compiled in approximately 50 species and distributed in approximately 40 structural types (Passaglia and Sheppard 2001).

However, this synthesis field still has areas that can be explored, including the search for possible new structures (Gilson 1992; Li et al. 2015). Computational studies have suggested the possibility of an even larger number of zeolitic structures (Morris and Čejka 2015); approximately 300,000 hypothetical structures for zeolites (and zeotypes) were calculated by Pophale et al. (2011) (Fig. 1.2).

Synthetic zeolites have gained much greater prominence and application in the contemporary world due to the greater selectivity and purity of synthetic zeolites compared with natural mineral zeolites, although modifying methods have been suggested to broaden the applications of natural zeolites (Ates and Akgül 2016).

Fig. 1.2 Energy density plot of known (red line) and hypothetical zeolite structures. Reproduced (adapted) with permission from Pophale et al. (2011), Copyright (2011) Royal Society of Chemistry

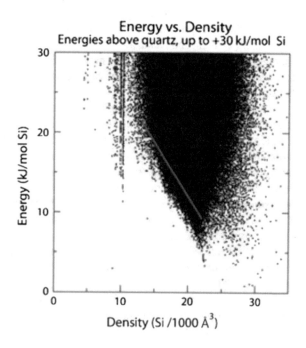

1.2 Concept

The search for the most coherent definition of a zeolite has continued throughout the research of these materials, since there are an increasing a number of materials that exhibit zeolite caracteristics, but different structural organizations and properties.

One "classic" definition found in many works is mineralogical, describing zeolites as a class of hydrated aluminosilicates of highly crystalline alkaline and alkaline earth metals with pores smaller than 2.0 nm. This definition is closely related to the geological origins of natural zeolites as minerals that are different from other aluminosilicates such as feldspars, feldspathoids, and scapolites[1] because of the more open structures and hydration of zeolites (Armbruster and Gunter 2001; Shams and Ahi 2013).

In 1986, Richard Barrer defined zeolites as tectosilicates, that is, three-dimensional anionic networks formed by the simultaneous union of SiO_4 and AlO_4 tetrahedra that share oxygen ions (Barrer 1986). However, Pastore (1996) argued that an aluminosilicate can only be considered zeolite if the material has more than one aluminum atom in each unit cell of its structure; otherwise, the material is considered to be a silicate doped with aluminum.

Using the idea of tectosilicates, Payra and Dutta (2003) presented a division made by Liebau et al. (1986) for porous tectosilicates in which zeolites are a distinct group of this class of aluminosilicates, together with the zeolite sodalite, called chlathralite. Pure silica zeolitic structures are classified as zeosils.

Over time, more comprehensive definitions have been proposed to incorporate materials with similar characteristics that have emerged. One definition proposed by the Subcommittee on Zeolites of the International Mineralogical Association of the Commission on New Minerals and Mineral Names considers every crystalline substance whose structure is characterized by a skeleton of interconnected tetrahedra, each consisting of four oxygen atoms and a cation, that form structures of channels and cavities and that can be interrupted by OH or F groups to be zeolites (Resende et al. 2008; Coombs et al. 1998).

Baerlocher et al. (2007) proposed that a three-dimensional network formed by atoms tetrahedrally coordinated to oxygen atoms, TO_4, with a structural density between 19 and 21 T atoms per 1000 Å^3 can be considered to be a zeolitic or zeotypic material.

According to these new concepts, molecular sieves such as aluminophosphates (AlPOs) and silicoaluminophosphates (SAPOs) can be classified as zeolites. In this sense, zeotype (zeolite-like) refers to materials whose structures and properties are similar to those of "traditional" zeolites.

A more recently reported definition described zeolites as silica-based microporous crystalline solids in which certain silicon atoms are replaced by other elements (generally denoted T) such as trivalent and tetravalent ions at a Si/T molar ratio lower

---

[1]Feldspars are more closed than zeolites, feldspathoids have a lower degree of aluminum in their structure, and scapolite differs by naturally possessing anionic structural groups. In addition, both groups consist of anhydrous minerals, while zeolites are hydrated (Klein and Dutrow 2012).

than 500 (Paillaud and Patarin 2016). The authors specified that zeolite materials with Si/T molar ratios greater than 500 are zeosils.

However, there is no consensus on a concise definition, and certain authors prefer to use more neutral definitions, referring to materials, for example, as crystalline networks based on silica or composed of covalent bonds (Muraoka et al. 2016; Anderson et al. 2017).

The concept of the molecular sieve, introduced by James W. McBain in 1932, as the selective adsorption property primarily refers to zeolites. This term commonly embraces any solid material with uniform porosity that exhibits selectivity in the adsorption of molecules (Howe-Grant 1998). The scope of this definition is wider since many nonzeolitic porous materials exhibit the ability to selectively adsorb.

Pastore (1996) used the term zeolitic metallosilicates for structures similar to zeolites that incorporate more than one T^{3+} or T^{4+} metal ion other than Al^{3+} per unit cell.

Finally, the IUPAC (McCusker 2005) has adopted a definition that covers microporous and mesoporous materials with inorganic hosts, including those with atypical zeolite chemical compositions, with nontetrahedral building units and/or host structures that do not extend in three dimensions, since there is accessible ordered porosity with free diameters smaller than 50 nm.

1.3   Zeolite LEGO: Structure and Taxonomy of Zeolites

Zeolites are solids with complex structures, although certain aspects are commented on below. Starting from the crystallographic unit cell of a zeolite, many papers adopt the general empirical formula (Eq. 1.1) of the composition as follows:

$$M\frac{y}{m}\left[(SiO_2)_x(AlO_2)_y\right] \cdot wH_2O \tag{1.1}$$

where M is the valence cation m—usually cations of the alkali metal (Group 1) or alkaline earth (Group 2) groups, although organic cations can be used—that counterbalances the negative charges in the zeolite structure, w is the number of water molecules contained in the unit cell, and x and y are the numbers of tetrahedra per unit cell.

Structurally, zeolites—treated as aluminosilicates—consist of an infinitely extended, continuous and rigid three-dimensional framework of silica tetrahedra $[SiO_4]^{4-}$ and alumina $[AlO_4]^{5-}$ connected by the oxygen ions (Ghobarkar et al. 2003). Each O^{2-} ion joins two tetrahedra by $T-O-T$ bonds with angles normally ranging from 140° to 165°, forming strands as a kind of infinite network inorganic polymer (Fig. 1.3) (Flanigen 2001; Flanigen et al. 2010).

Due to the trivalence of the aluminum ions, there is an imbalance of charges, giving the structure a negative charge that is usually compensated by metallic cations such as Na^+ and Ca^{2+}. The principle of "aluminum avoidance" or Lowenstein's rule,

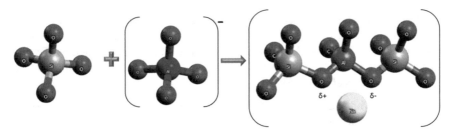

Fig. 1.3 Basic building units of zeolites. Silicon (gray)-oxygen (red)-aluminum (blue) bonds generate a negative charge, which is balanced by compensation cations (yellow sphere) between tetrahedral chains [Figure generated with Avogadro software (Hanwell et al. 2012)]

proposed by Lowenstein and Lowenstein (1954), describes two postulates. The first postulate states that whenever two tetrahedra are connected by an oxygen bridge, the central ion of one tetrahedron can be Al^{3+}, while the other is a tetravalent ion, notably Si^{4+}. The second postulate states that whenever two ions Al^{3+} are neighbors of the same O^{2-} anion, at least one of the Al^{3+} ions must have a coordination number above 4.

Ab initio calculations show that $Al-O-Al$ bonds are less stable and are therefore more unfavorable in zeolites prepared by conventional synthetic routes (Bell et al. 1992; Larin 2013), although exceptions have been reported (Fletcher et al. 2017).

These rules mean that for zeolitic aluminosilicates, the Si/Al ratio can assume a maximum value of 50% of the Si^{4+} ions substituted by Al^{3+} ions, and together with the ratios of O/Si and O/Al, there may be twice as many O^{4-} ions as T^{x+} ions. The possibility of tetrahedra formed by oxides of other elements, such as $[TiO_4]$ and $[PO_4]^{3-}$, in zeolite structures was raised (Man and Sauer 1996; Ghobakar et al. 2003), or oxygen be replaced by sulfur, for example (Pastore 1996).

$[SiO_4]^{4-}$ and $[AlO_4]^{5-}$ tetrahedra are arranged in the form of rings that combine to form regular and uniform channels and cavities. Water molecules with freedom of movement are found in the channels in addition to the compensating cations such as Na^+, Ca^{2+}, Mg^{2+} and K^+ (Giannetto et al. 2000). These channels and well-defined molecular cavities give zeolites microporous structures with much larger inner surfaces than outer surfaces (Chatterjee 2010).

The greatest overall difference between zeolites and mesoporous sieves or other porous materials, such as coal, is the extremely organized arrangement of the tetrahedra. The structural organization of molecular sieves provides pores, cavities and channels with uniform sizes, giving the molecular sieves high selectivity regarding the chemical species that enter, exit and settle in their channels.

The zeolite pores are sized (0.3–1.3 nm or 3.0–13 Å) very similarly to the kinetic diameters of a large number of molecules (Ghosh et al. 2009). The aperture of a pore is described by the number of tetrahedrally coordinated elements connected in sequence that circumscribe the window of the member rings (MR). Thus, a pore defined by a ring of 12 T atoms is written as 12-MR (Balkus 2002). Compensation cations also affect the pore openings.

The three-dimensional interactions lead to many different geometries, from large internal cavities to a network of channels that cross the entire zeolitic crystal (Braga and Morgon 2007). Thus, IUPAC established pore size limits by dividing the pore sizes into three general categories using an adsorption-desorption technique with N_2 at 77 K (Thommes et al. 2015):

- Pores with widths greater than 50 nm are classified as macropores. Many ceramic materials and metal oxide films are in this group.
- Pores with widths between 2.0 and 50 nm are classified as mesopores. Sieves such as MCM[2]-41, SBA[3]-16 and KIT[4]-6 are in this group.
- Pores smaller than 2.0 nm are micropores. Most zeolites fit in this group. Na-LTA zeolite, for example, has approximately cylindrical pores of 0.41 nm or 4.1 Å in diameter, and mordenite has elliptical pores sized 6.5 × 7.0 Å (Baerlocher et al. 2007).

In addition, micropores sized approximately 0.7 nm in width have recently been divided into ultramicropores, and those between 0.7 nm and 2.0 nm in width are called supermicropores (Thommes et al. 2015).

The pore diameter is another way of classifying molecular sieves, as in Table 1.1.

The TO_4 tetrahedra are understood to be the basic building units (BBUs) of zeolites. However, paraphrasing Barrer et al. (1959), it is highly difficult to conceive the construction of continuous and complex spatial structures solely by the connection of tetrahedra and relate the structure with the symmetry of its unit cell. Therefore, Barrer et al. (1959) proposed a rationalization of the mechanism of silicate species formation in solution through the concept of secondary building units (SBUs) (Fig. 1.4).

According to Smith (2000), the physical chemistry of zeolites is based on mathematical principles of 3D geometry. To facilitate the description of structures, researchers such as Breck (1974) developed and introduced the use of SBUs in the structural classification of zeolites since the Barrer proposal was widely accepted a priori even by the IZA (Knight 1990; Anurova et al. 2010).

Table 1.1 Pore diameter classification for molecular sieves

Type	O atoms on ring pore	Diameter (Å)	Example	Dimensionality[a]
Extra large	>12	$9 < \mathbf{d} < 20$	IFO, AEI	1, 3
Large	12	$6 < \mathbf{d} < 9$	MOR, β, FAU	2, 3
Medium	10	$4.5 < \mathbf{d} < 6$	MFI, MEL	3, 1
Small	8	$3 < \mathbf{d} < 4.5$	LTA	3

[a]1 = 1D channel, 2 = 2D channel and 3 = 3D channel
Source Li and Yu (2014), Baerlocher and McCusker 2018

---

[2]Abbreviation for "Mobil Composition of Matter".

[3]Abbreviation for "Santa Barbara Amorphous".

[4]Abbreviation for "Korea Advanced Institute of Science and Technology".

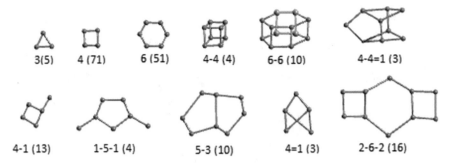

Fig. 1.4 Certain secondary building units. The numbering below each unit refers to the number of tetrahedra as they are bound. The numbers in parentheses refer to the occurrence

SBUs are idealized structural description elements of composite anions containing up to 16 T atoms resulting in polygons and polyhedra, such as single (S#R[5]) and double (D#R) rings, representing the different ways in which TO_4 tetrahedra may form composite arrangements. In turn, when combined in an integer number of units, SBUs can develop the various unit cells characteristic of each zeolite (Barrer et al. 1959; Baerlocher et al. 2007).

The SBU concept is very useful for understanding the topology of zeolites. However, its application has been questioned (Knight 1990), because there is no experimental evidence to support the hypothesis that zeolite growth occurs by sequential addition of SBUs rather than simple silicate species such as monomers and dimers. Nevertheless, SBUs remain in use as tools for intellection of structures (Mortier and Schoonheydt 1985; Smith 2000).

In addition to the SBUs, other finite units were designed for the taxonomic study of zeolites. These forms are collectively called polyhedral building units (PBUs) or structural subunits (SSUs) (Fig. 1.5). PBUs are more abstruse units that appear easily when analyzing the cavity or pore system of a zeolite—such as the α-cavity—but can be thought of as arising from combinations of SBUs. These structures can be applied together with SBUs to provide better understanding of crystal growth in synthesis (Koningsveld 1991; Braga and Morgon 2007).

In his Compendium of Zeolite Framework Types, Koningsveld (2007) describes several zeolite structures with what he called periodic building units (PerBUs) that differ in part from SBUs in that PerBUs encompass finite units (simple and double rings) and infinite units (chains, tubes and knits)—the fundamental building units (FBUs)—proposed by McCusker et al. (2001) and are combined through the smallest number of connections and symmetry operations such as translation, rotation and reflection (Braga and Morgon 2007).

These units stand out because, in addition to their varied elements, these units are used in a document entitled "Schemes of Zeolite Framework Types", available on

---

[5]Codes for mentioning SBU types. "S" stands for "simple", "D" is equivalent to "double", "R" is "ring" and "#" is the number of tetrahedra involved in the unit..

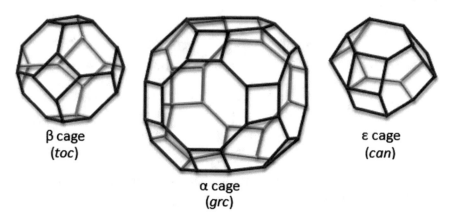

β cage
(*toc*)

ε cage
(*can*)

α cage
(*grc*)

Fig. 1.5 Certain better known PBUs. The codes in parentheses were established by Smith (2000) to identify the various PBUs proposed by the same

the IZA Structure Commission website, that describes the structures of the zeolites found in the online collection of the association Baerlocher and McCusker 2018.

Sometimes, in addition to unitary polyhedric units, smaller unit chains or even 2D networks can be used following the same idea of stacking beads for the assembly of single unit cells (McCusker and Baerlocher 2005; Xu et al. 2007).

This accumulation of concepts is due to an absence of rules organizing the construction unit systems and can cause misunderstandings in zeolite structural descriptions. However, Anurova et al. (2010) give a general overview on the subject and explain that a hierarchy of these concepts has been developed (Fig. 1.6) in which all the aforementioned construction units are connected and belong to a system of composition buildings units CBUs. CBUs are therefore a grouping of finite and infinite building units and involve various forms, such as SBUs, polyhedra such as the sodalite unit (β cavity) and chains of tetrahedra (McCusker et al. 2001; Baerlocher et al. 2007).

In view of the problems discussed, a crystal structure description concept called natural tiling has been proposed. The term tiling comes from mathematics and is defined as a periodic division of Euclidean space that can be used to cover a plane; the use of tiling to describe zeolites has been supported by computational calculations (Delgado-Friedrichs et al. 1999; Blatov et al. 2007).

Tilings exist in large numbers; however, Blatov et al. (2007), proposed a set of rules for zeolites that creates the natural units of construction or natural tilings (Fig. 1.7) (Anurova et al. 2010).

Natural building units (NBUs) promote unequivocal division of space and always conform to the smallest pore of the structure, and other cavities can be constructed as sums of different NBUs joined by one or more faces. In addition, any network window matches one face of an NBU, and therefore, NBUs correspond to the actual cavities, boxes and channels. Finally, NBUs suitable for a given structure can be

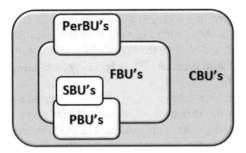

Fig. 1.6 Hierarchical relationships between types of construction units: secondary building units (SBUs) and polyhedral building units (PBUs) are basically encompassed in the fundamental building units (FBUs). This group together with the periodic building units (PerBUs) is the more general group of composite building units (CBUs). Reproduced (adapted) with permission from Anurova et al. (2010), Copyright (2010) American Chemical Society

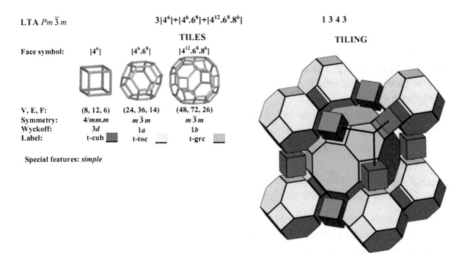

Fig. 1.7 Representation of the NBUs for the LTA structure. Reproduced (adapted) with permission from Anurova et al. (2010), Copyright (2010) American Chemical Society

determined by computational methods and can promote a universal classification scheme (Anurova et al. 2010).

The group of zeolitic materials is vast, so the same structure can aggregate different chemical compositions, resulting in various zeolites and zeotypes. For example, the MFI structure, when composed of Si and Al, is encoded as ZSM-5, and the same framework also describes silicalite-1 (pure Si) and TS-1 (Si and Ti). Given these differences, methods of classification and organization must be adopted. In this sense, the IZA adopted systematic topological representation through a mnemonic code of three capital letters for each zeolitic structure (Smith 2000).

The codes are related to the zeolite first identified with the structure. The codes can relate to natural zeolites, such faujasite with the code FAU, which is also the structure of zeolites X and Y, or to the organization that developed the zeolite, such as zeolite A and ZK-4 with the code LTA, originating from Linde, a division of Union Carbide (Barrer 1979).

Based on the structure topology, Breck (1974) suggested a classification scheme for zeolites in which the structures were divided into seven groups based on the familiarity of the building units. This scheme was modified by Glauco Gottardi and Ermanno Galli to include the historical contexts of the findings and is widely used by geologists (Table 1.2) (Armbruster and Gunter 2001).

To ensure researcher understanding of the structural descriptions of porous materials, the chemist and crystallographer Lynne Bridget McCusker, together with collaborators, established IUPAC classifications for the various types of openings found in zeolites that can be considered to be the components that form the pores (McCusker et al. 2001):

- Windows: TO_4 unit rings that define the faces of a polyhedral pore, for example, an SBU S4R;
- Cages: polyhedral pores in which the windows are too narrow to allow the passage of molecules larger than water molecules, for example, the sodalite unit, a truncated octahedron;
- Cavities: polyhedral pores with at least one face defined by a ring large enough to be penetrated by a host species but not infinitely extensive. These elements have larger internal spaces than the input. For example, an α-cavity, a truncated cuboctahedron;
- Channels: pores extending infinitely in one dimension and wide enough to allow the diffusion of molecules along their length.

The nature of the channels and cavities formed in the zeolitic structure when dehydrated is important in determining their physical and chemical properties. Channel systems are classified into the following three types (Breck 1974):

Table 1.2 Zeolite structural groups

Group	Secondary building unit	Structure (Ex.)
1	Single 4-ring tetrahedra (S4R)	Chabasite
2	Single/double 6-ring tetrahedra (S6R/D6R)	Offretite/Faujasite
3	Double 4-ring tetrahedra (D4R)	LTA
4	Fibrous (4–1) $[T_5O_{10}]$	Natrolite
5	Mordenite group (5–1) $[T_8O_{16}]$	Mordenite
6	Heulandite group (4–4–1) $[T_{10}O_{20}]$	Heulandite
7	Unknown Structures	Cowlesite

Source Baerlocher and McCusker 2018

(a) One-dimensional: the channels do not intersect—(L, AlPO4-5);
(b) Two-dimensional: two types of channels, with the same sizes or not, intersect—(mordenite, ZSM-5, ferrierite);
(c) Three-dimensional: channels, with the same dimensions or not, intersect in three directions—[faujasite (X and Y), LTA].

Zeolite structures give materials a number of physical properties, such as: low density (1.8–2.3 g cm^{-3}), mechanical hardness in the range of 4–6 on the Mohs scale, refractive indexes generally varying between 1.47 and 1.52 (Poole 2004), ion exchange capacity, and several other physical and chemical properties that allow the applications of these materials in many uses.

1.4 Natural and Synthetic Zeolites

The natural zeolites (Table 1.3) are aluminosilicate ores formed by isomorphic substitution in the silicates and are found in mafic volcanic rocks of basaltic lava, deposits of saline, alkaline environments, and diagenetic, hydrothermal and marine sediments (Colella 2005; Resende et al. 2008). In Brazil, for example, there are three natural zeolite deposits (Fig. 1.8).

Pure zeolites, like other minerals, are colorless or white, acquiring coloration from traces of certain transition metals (Ghobakar et al. 2003).

Minerals are formed in various environmental conditions of pressure, temperature, ionic species activity and partial pressure of water among other factors. Consequently, considerable variations in chemical composition are expected to occur.

Natural zeolites are commonly used in functions that do not require a high degree of purity or application specificity, for example, animal nutrition (Papaioannou et al. 2005), loading materials for the paper industry, odor control, treatment of radioactive liqueurs, water and effluent treatment, detergents, desiccation, solar heating, refrigeration, rock blocks, pozzolanic cement and light aggregates for civil construction

Table 1.3 Certain natural zeolites and their unit cell formations

Zeolite	Chemical formula
Faujasite	$\left|(Ca^{2+}, Mg^{2+}Na_2^+)_{29}(H_2O)_{240}\right|[Al_{58}Si_{134}O_{384}]$
Chabazite	$\left|Ca_6^{2+}(H_2O)_{40}\right|[Al_{12}Si_{24}O_{72}]$
Mordenite	$\left|Na_8^+(H_2O)_{24}\right|[Al_8Si_{40}O_{96}]$
Analcime	$\left|Na_{16}^+(H_2O)_{16}\right|[Al_{16}Si_{32}O_{96}]$
Heulandite	$\left|Ca_4^{2+}(H_2O)_{24}\right|[Al_8Si_{28}O_{72}]$

Source Baerlocher and McCusker 2018

	PARNAÍBA BASIN 1. Stilbite and laumontite on Corda Formation (MA/TO States)
	POTIGUAR BASIN 2. Philipsite, harmotome and analcime on Macau Formation (RN State)
	PARANA BASIN 3. Philipsite on Uberaba Formation (MG State) 4. Analcime on Adamantina Formation (SP State) 5. Heulandite on Batucatu Formation (MS State)

Fig. 1.8 Zeolites on sedimentary rocks in Brazil. Reproduced (adapted) with permission from Luz AB, Lins FAF. Rochas & Minerais Industriais: Usos e especificações, 2nd edn. CETEM/MCT, Rio de Janeiro

(Mumpton 1999), gas purification and separation (Ackley 2003) and agriculture and agrochemistry (Reháková et al. 2004).

Currently, more than 200 zeolitic topologies are known, most of which are obtained only in the laboratory. Natural zeolites have problems with impurities and possible composition differences even on a single extraction shaft. However, synthetic zeolites present high degrees of purity, being monocationic or dicationic, and thermal stability, mainly when the compensation cation is an alkaline-earth metal (Resende et al. 2008).

The isomorphic substitution of Si and Al atoms by atoms such P (Fatourehchi et al. 2011), B (Qiao et al. 2014) and Ti (Deng et al. 2013), generating other types of molecular sieves or zeotypes, is a process that can be manipulated to produce specific catalysts (Pastore 1996;).

Although it is difficult to estimate the total world production of zeolites due to the application of natural zeolites, in 2009, the world production of synthetic zeolites was estimated to be 1.7–2 million Mg year^{-1}, (1.0 Mg = 10^3 kg), and that of natural zeolites was approximately 2 billion Mg year^{-1} (Yilmaz and Müller 2009), most of which was destined for the production of detergents with approximately 1/3 to catalytic processes, mainly in petrochemicals.

The importance of natural zeolites is undeniable. However, due to the properties, the possibility of greater control and design, the exclusion of impurities, the possibility of more refined applications and even quantitative differences, synthetic zeolites are plausibly more prominent and applicable in industries, mainly in the production of catalysts.

References

Ackley M (2003) Application of natural zeolites in the purification and separation of gases. Microporous Mesoporous Mater 61:25–42. https://doi.org/10.1016/s1387-1811(03)00353-6

Anderson MW, Gebbie-Rayet JT, Hill AR et al (2017) Predicting crystal growth via a unified kinetic three-dimensional partition model. Nature 544:456–459. https://doi.org/10.1038/nature21684

Anurova NA, Blatov VA, Ilyushin GD, Proserpio DM (2010) Natural tilings for zeolite-type frameworks. J Phys Chem C 114:10160–10170. https://doi.org/10.1021/jp1030027

Argauer RJ, Landolt GR (1972) Crystalline zeolite ZSM-5 and method of preparing the same. US Patent 3,702,886, 14 Nov 1972

Armbruster T, Gunter ME (2001) Crystal structures of natural zeolites. In: Bish DL, Ming DW (eds) Natural zeolites: occurrence, properties and applications. Reviews in mineralogy and geochesmitry, vol 45. Mineralogical Society of America, Virginia, pp 1–67

Ates A, Akgül G (2016) Modification of natural zeolite with NaOH for removal of manganese in drinking water. Powder Technol 287:285–291. https://doi.org/10.1016/j.powtec.2015.10.021

Baerlocher C, McCusker LB (2018) Database of zeolite structures. http://www.iza-structure.org/databases/. Accessed 26 Oct 2018

Baerlocher C, McCusker LB, Olson DH (2007) Atlas of zeolite framework types, 6th edn. Elsevier, Amsterdã, p 3

Balkus KJ (2002) Synthesis of large pore zeolites and molecular sieves. In: Karlin KD (ed) Progress in inorganic chemistry. Progress in inorganic chemistry series, vol 50. Wiley, New York, pp 217–268

Barrer RM (1948) 435. Syntheses and reactions of mordenite. J Chem Soc (Resumed): 2158

Barrer RM (1979) Chemical nomenclature and formulation of compositions of synthetic and natural zeolites. Pure Appl Chem 51:1091–1100. https://doi.org/10.1351/pac197951051091

Barrer RM (1986) Zeolite synthesis: an overview. In: Basset J-M, Gates BC, Candy J-P, Choplin A, Leconte M, Quignard F, Santini C (eds) Surface organometallic chemistry: molecular approaches to surface catalysis. Kluwer Academic Publishers, France, pp 221–243

Barrer RM, Baynham JW, Bultitude FW, Meier WM (1959) 36. Hydrothermal chemistry of the silicates. Part VIII. Low-temperature crystal growth of aluminosilicates, and of some gallium and germanium analogues. J Chem Soc (Resumed) 195. https://doi.org/10.1039/jr9590000195

Bell RG, Jackson RA, Catlow CRA (1992) Löwestein's rule in zeolite A: a computational study. Zeolites 12:870–871

Bieseki L, Simancas R, Jordá JL et al (2018) Synthesis and structure determination via ultra-fast electron diffraction of the new microporous zeolitic germanosilicate ITQ-62. Chem Commun 54:2122–2125. https://doi.org/10.1039/c7cc09240g

Blatov VA, Delgado-Friedrichs O, O'keeffe M et al (2007) Periodic nets and tilings: possibilities for analysis and design of porous materials. In: Xu R, Gao Z, Chen J et al (eds) From zeolites to porous MOF materials—The 40th anniversary of International Zeolite Conference. Proceedings of 15th International Zeolite Conference, Beijing, 12–17 August (Studies in surface science and catalysis), vol 170B. Elsevier, Amsterdã, pp 1637–1645

Braga ACA, Morgon NH (2007) Descrições estruturais cristalinas de zeólitos. Quim Nova 30(1):178–188

Breck DW (1964a) Crystalline molecular sieves. J Chem Educ 41(12):678

Breck DW (1964b) Crystalline zeolite Y. US Patent 3,130,007, 21 Apr 1964

Breck DW (1974) Zeolite molecular sieves. Structure, chemistry, and use. Wiley, New York, p 771

Breck DW, Eversole WG, Milton RM et al (1956) Crystalline zeolites. I. The properties of a new synthetic zeolite, type A. J Am Chem Soc 78:5963–5972. https://doi.org/10.1021/ja01604a001

Byrappa K, Yoshimura M (2013) Physical Chemistry of hydrothermal growth of crystals. In: Byrappa K, Yoshimura M (eds) Handbook of hydrothermal technology, 2nd edn. Elsevier, Oxford, pp 139–175

Camblor MA, Corma A, Díaz-Cabañas M-J, Baerlocher C (1998) Synthesis and structural characterization of MWW type zeolite ITQ-1, the pure silica analog of MCM-22 and SSZ-25. J Phys Chem B 102:44–51. https://doi.org/10.1021/jp972319k

Chatterjee A (2010) Structure property correlations for nanoporous materials. CRC Press, New York, p 183

Colella C (2005) Natural zeolites. In: Čejka J, Van Bekkum H (eds) Zeolites and ordered mesoporous materials: progress and prospects. In: 1st FEZA School of zeolite, Prague, August 2005 (Studies in Surface Science and Catalysis), vol 157. Elsevier, Amsterdã, pp 13–40

Coombs DS, Alberti A, Armbruster T et al (1998) Recommended nomenclature for zeolite minerals: report of the subcommittee on zeolites of the International Mineralogical Association, Commission on New Minerals and Mineral Names. Mineral Mag 62:533–571. https://doi.org/10.1180/002646198547800

Corma A, Fornes V, Pergher SB et al (1998) Delaminated zeolite precursors as selective acidic catalysts. Nature 396:353–356. https://doi.org/10.1038/24592

Corrêa MLS, Wallau M, Schuchardt U (1996) Zeólitas tipo AlPO: Síntese, caracterização e propriedades catalíticas. Química Nova, [S.I], vol 19, n 1, pp 43–50.

Cundy CS, Cox PA (2003) The hydrothermal synthesis of zeolites: history and development from the earliest days to the present time. Chem Rev 103:663–702. https://doi.org/10.1021/cr020060i

Cundy CS, Cox PA (2005) The hydrothermal synthesis of zeolites: precursors, intermediates and reaction mechanism. Microporous Mesoporous Mater 82(1–2):1–78

Delgado-Friedrichs O, Dress AWM, Huson DH et al (1999) Systematic enumeration of crystalline networks. Nature 400:644–647. https://doi.org/10.1038/23210

Deng X, Wang Y, Shen L et al (2013) Low-cost synthesis of titanium silicalite-1 (TS-1) with highly catalytic oxidation performance through a controlled hydrolysis process. Ind Eng Chem Res 52:1190–1196. https://doi.org/10.1021/ie302467t

Estermann M, Mccusker LB, Baerlocher C et al (1991) A synthetic gallophosphate molecular sieve with a 20-tetrahedral-atom pore opening. Nature 352:320–323. https://doi.org/10.1038/352320a0

Fatourehchi N, Sohrabi M, Royaee SJ, Mirarefin SM (2011) Preparation of SAPO-34 catalyst and presentation of a kinetic model for methanol to olefin process (MTO). Chem Eng Res Des 89:811–816. https://doi.org/10.1016/j.cherd.2010.10.007

Flanigen EM, Patton RL (1978) Silica polymorph and process for preparing the same. US Patent 4,073,865, 14 Feb 1978

Flanigen EM (2001) Zeolites and Molecular Sieves: An historical perspective. In: Van Bekkum H, Flanigen EM, Jansen KJC (eds). Introduction to Zeolite Science and Practice. 2. ed Amsterdã: Elsevier B.V. Cap. 2. pp 11–35. (Studies in Surface Science and Catalysis. vol 137)

Flanigen EM, Broach RW, Wilson ST (2010) Introduction. In: Kulprathipanja S (ed) Zeolites in industrial separation and catalysis, 1st edn. Wiley-VCH, Weinheim, pp 1–26

Fletcher RE, Ling S, Slater B (2017) Violations of Löwensteins rule in zeolites. Chem Sci 8:7483–7491. https://doi.org/10.1039/c7sc02531a

Ghobakar H, Schäf O, Massiani Y, Knauth P (2003) The reconstruction of natural zeolites, 1st edn. Kluwen Academic Publishers, Netherlands, p 1

Ghosh A, Jordan E, Shantz DF (2009) Applications of microporous and mesoporous materials. In: Klabunde KJ, Richards RM (eds) Nanoscale materials in chemistry, 2nd edn. Wiley, New Jersey, pp 331–366

Giannetto GP, Montes AR, Rodriguéz GF (2000) Zeolitas: características, propiedades y aplicaciones industriales. Editorial Innovacíon Tecnológica, Facultad de Ingeniería, UCV, Caracas

Gilson J-P (1992) Organic and inorganic agents in the synthesis of molecular sieves. In: Zeolite microporous solids: synthesis, structure, and reactivity, pp 19–48. https://doi.org/10.1007/978-94-011-2604-5_2

Gottardi G, Galli E (1985) Zeolites with 6-Rings. In: Gottardi G, Galli E (eds) Natural zeolites. Minerals and rocks series, vol 18. Springer, Heidelberg, pp 168–222

Guisnet M, Ribeiro FR (2004) Zeólitos: Um nanomundo ao serviço da catálise. Lisboa: Fundação Calouste Gulbenkian. p 221

Hanwell MD, Curtis DE, Lonie DC et al (2012) Avogadro: an advanced semantic chemical editor, visualization, and analysis platform. J Cheminform 4:17. https://doi.org/10.1186/1758-2946-4-17

Howe-Grant M (1998) Kirk-Othmer encyclopedia of chemical technology, vol 16, 4th edn. Wiley

Kerr GT (1966) Chemistry of crystalline aluminosilicates. II. The synthesis and properties of zeolite ZK-4. Inorg Chem 5:1537–1539. https://doi.org/10.1021/ic50043a015

Knight CTG (1990) Are zeolite secondary building units really red herrings? Zeolites 10:140–144. https://doi.org/10.1016/0144-2449(90)90036-q

Klein C, Dutrow B (2012) Manual of mineral science, 23th edn. Wiley, USA. Portuguese edition: Klein C, Dutrow B (2012) Manual de Ciência dos Minerais (trans: Menegat R). Bookman, São Paulo

Koningsveld HV (1991) Structural subunits in silicate and phosphate structures. In: Bekkum HV, Flaingen EM, Jansen JC (eds) Introduction to zeolite science and practice. Studies in surface science and catalysis, vol 58. Elsevier, Amsterdã, pp 35–76

Koningsveld HV (2007) Compendium of zeolite framework types: building schemes and type characteristics. Elsevier, Amsterdã

Larin AV (2013) The Loewenstein rule: the increase in electron kinetic energy as the reason for instability of Al–O–Al linkage in aluminosilicate zeolites. Phys Chem Miner 40:771–780. https://doi.org/10.1007/s00269-013-0611-7

Lazlo P (2018) Two laboratory deaths, and keeping organic solvents dry. Angew Chem Int Ed 57:8822–8824. https://doi.org/10.1002/anie.201803276

Li Y, Yu J (2014) New stories of zeolite structures: their descriptions, determinations, predictions, and evaluations. Chem Rev 114:7268–7316. https://doi.org/10.1021/cr500010r

Li J, Corma A, Yu J (2015) Synthesis of new zeolite structures. Chem Soc Rev 44:7112–7127. https://doi.org/10.1039/c5cs00023h

Liebau F, Gies H, Gunawardane R, Marler B (1986) Classification of tectosilicates and systematic nomenclature of clathrate type tectosilicates: a proposal. Zeolites 6:373–377. https://doi.org/10.1016/0144-2449(86)90065-5

Lok BM, Messina CA, Patton RL et al (1984) Silicoaluminophosphate molecular sieves: another new class of microporous crystalline inorganic solids. J Am Chem Soc 106:6092–6093. https://doi.org/10.1021/ja00332a063

Lowenstein W, Lowenstein M (1954) The distribution of aluminum in the tetrahedra of silicates and aluminates. Am Mineral 39:92–96

Man AJMD, Sauer J (1996) Coordination, structure, and vibrational spectra of titanium in silicates and zeolites in comparison with related molecules. An ab initio study. J Phys Chem 100:5025–5034. https://doi.org/10.1021/jp952790i

Masters AF, Maschmeyer T (2011) Zeolites—from curiosity to cornerstone. Microporous Mesoporous Mater 142:423–438. https://doi.org/10.1016/j.micromeso.2010.12.026

McCusker LB (2005) IUPAC nomenclature for ordered microporous and mesoporous materials and its application to non-zeolite microporous mineral phases. Rev Mineral Geochem 57:1–16. https://doi.org/10.2138/rmg.2005.57.1

McCusker LB, Baerlocher C (2005) Zeolite structures. In: Čejka J, Bekkum HV (eds) Zeolites and ordered mesoporous materials: progress and prospects. Studies in surface science and catalysis, vol 157, 1st edn. Elsevier, Amsterdã, pp 41–64

McCusker LB, Liebau F, Engelhardt G (2001) Nomenclature of structural and compositional characteristics of ordered microporous and mesoporous materials with inorganic hosts (IUPAC recommendations 2001). Pure Appl Chem 73:381–394. https://doi.org/10.1351/pac200173020381

McDaniel CV, Maher PK (1969) Stabilized zeolites. US Patent 3,449,070, 10 June 1969

Millini R, Belussi G (2017) Zeolite science and perpectives. In: Čejka J, Morris R, Nachtigall P (eds) Zeolites in catalysis: properties and applications. RSC catalysis series, vol 28. The Royal Society of Chemistry, Cryondon, pp 1–36

Mintova S, Grand J, Valtchev V (2016) Nanosized zeolites: quo vadis? C R Chim 19:183–191. https://doi.org/10.1016/j.crci.2015.11.005

Möller K, Bein T (2011) Pores within pores—how to craft ordered hierarchical zeolites. Science 333:297–298. https://doi.org/10.1126/science.1208528

Morris RE, Čejka J (2015) Exploiting chemically selective weakness in solids as a route to new porous materials. Nat Chem 7:381–388. https://doi.org/10.1038/nchem.2222

Mortier WJ, Schoonheydt RA (1985) Surface and solid state chemistry of zeolites. Prog Solid State Chem 16:1–125. https://doi.org/10.1016/0079-6786(85)90002-0

Mumpton FA (1999) La roca magica: uses of natural zeolites in agriculture and industry. Proc Natl Acad Sci 96:3463–3470. https://doi.org/10.1073/pnas.96.7.3463

Muraoka K, Chaikittisilp W, Okubo T (2016) Energy analysis of aluminosilicate zeolites with comprehensive ranges of framework topologies, chemical compositions, and aluminum distributions. J Am Chem Soc 138:6184–6193. https://doi.org/10.1021/jacs.6b01341

Paillaud J-L, Patarin J (2016). Initial materials for synthesis of zeolites. In: Mintova S (ed) Verified syntheses of zeolitic materials, 3rd edn. Synthesis Commission of International Zeolite Association, pp 24–28

Papaioannou D, Katsoulos P, Panousis N, Karatzias H (2005) The role of natural and synthetic zeolites as feed additives on the prevention and/or the treatment of certain farm animal diseases: a review. Microporous Mesoporous Mater 84:161–170. https://doi.org/10.1016/j.micromeso.2005.05.030

Passaglia E, Sheppard RA (2001) The crystal chemistry of zeolites. Rev Mineral Geochem 45:69–116. https://doi.org/10.2138/rmg.2001.45.2

Pastore HO (1996) A lógica da substituição isomórfica em peneiras moleculares. Quim Nova 19(4):372–376

Payra P, Dutta PK (2003) Zeolites: a primer. In: Auerbach SM, Carrado KA, Dutta PK (eds) Handbook of zeolite science and technology. Marcel Dekker, Inc., New York, pp 125–199

Poole CP Jr (2004) Encyclopedic dictionary of condensed matter physics, vol 1A-M. Elsevier, San Diego, p 1507

Pophale R, Cheeseman PA, Deem MW (2011) A database of new zeolite-like materials. Phys Chem Chem Phys 13:12407. https://doi.org/10.1039/c0cp02255a

Qiao Q, Wang R, Gou M, Yang X (2014) Catalytic performance of boron and aluminium incorporated ZSM-5 zeolites for isomerization of styrene oxide to phenylacetaldehyde. Microporous Mesoporous Mater 195:250–257. https://doi.org/10.1016/j.micromeso.2014.04.042

Reháková M, Čuvanová S, Dzivák M et al (2004) Agricultural and agrochemical uses of natural zeolite of the clinoptilolite type. Curr Opin Solid State Mater Sci 8:397–404. https://doi.org/10.1016/j.cossms.2005.04.004

Resende NGAM, Monte MBM, Paiva PRP (2008) Zeólitas naturais. In: Luz, AB, Lins FAF (eds) Rochas & Minerais Industriais: Usos e especificações, 2nd edn. CETEM/MCT, Rio de Janeiro, pp 889–915

Shams K, Ahi H (2013) Synthesis of 5A zeolite nanocrystals using kaolin via nanoemulsion-ultrasonic technique and study of its sorption using a known kerosene cut. Microporous Mesoporous Mater 180:61–70. https://doi.org/10.1016/j.micromeso.2013.06.019

Smith JV (2000) Tetrahedral frameworks of zeolites, clathrates and related materials. Physical chemistry—Landolt-Börnstein: numerical data and functional relationships in science and technology, vol 14A. Springer, New York, p 266

Stocker K, Ellersdorfer M, Lehner M, Raith JG (2017) Characterization and utilization of natural zeolites in technical applications. BHM Berg- Huettenmaenn Monatsh 162:142–147. https://doi.org/10.1007/s00501-017-0596-5

Szostak R (1992) Handbook of molecular sieves. Van Nostrand Reinhold, New York, p 343

Taramasso M, Perego G, Notari B (1983) Preparation of porous crystalline synthetic material comprised of silicon and titanium oxides. US Patent 4,410,501, 18 Oct 1983

Thommes M, Kaneko K, Neimark AV et al (2015) Physisorption of gases, with special reference to the evaluation of surface area and pore size distribution (IUPAC technical report). Pure Appl Chem. https://doi.org/10.1515/pac-2014-1117

Wadlinger RL, Kerr GT, Rosinski EJ (1967) Catalytic composition of a crystalline zeolite. US 3,308,069, 7 Mar 1967

Weigel O, Steinhoff E (1924) IX. Die Aufnahme organischer Flüssigkeitsdämpfe durch Chabasit. Z Kristallogr-Cryst Mater 61:125–164

Wilson ST, Lok BM, Messina CA, Cannan TR, Flanigen EM (1983) Aluminophosphate molecular sieves: a new class of microporous crystalline inorganic solids. In: Stucky GD, Dwyer FG (eds) Intrazeolite chemistry. ACS symposium series, vol 218. American Chemical Society, Washington DC, pp 79–106

Xu R, Pang W, Yu J, Huo Q et al (2007) Chemistry of zeolites and related porous materials: synthesis and structure. Wiley, Singapore

Yamane I, Nakazawa T (1986) Development of zeolite for non-phosphated detergents in Japan. Pure Appl Chem 58:1397–1404. https://doi.org/10.1351/pac198658101397

Yilmaz B, Müller U (2009) Catalytic applications of zeolites in chemical industry. Top Catal 52:888–895. https://doi.org/10.1007/s11244-009-9226-0

Zimmermann NER, Haranczyk M (2016) History and utility of zeolite framework-type discovery from a data-science perspective. Cryst Growth Des 16:3043–3048. https://doi.org/10.1021/acs.cgd.6b00272

Chapter 2
Zeolite Synthesis: General Aspects

Studies focusing on zeolite synthesis in the field of sol-gel chemistry have been carried out since the 1940s. The large number of studies on this subject is due to both the scientific interest in the complexity and diversity of molecular sieves and the potential applicability of nanoporous solids in industrial processes (Livage 1994; Cundy and Cox 2005).

In this context, not only is the achievement of new structures and chemical compositions and improved synthesis routes desired but also a better understanding of the formation mechanisms and the roles of components and precursors, since these topics imply greater control over the synthesis.

2.1 Synthesis Methods

The most common method for synthesis of zeolites follows the hydrothermal route (Cundy and Cox 2005). Byrappa et al. (2015) use this term to refer to any chemical reaction with aqueous or nonaqueous solvents at temperatures and pressures above ambient conditions in a closed system that dissolves and recrystallizes relatively insoluble materials. Hydrothermal conditions can intensify hydrolysis and accelerate reaction rates according to the Arrhenius ratio, where the constant velocity of a reaction increases exponentially with temperature according to Eq. (2.1) (Xu et al. 2007) as follows:

$$\frac{d \ln k}{dT} = \frac{E}{RT^2} \tag{2.1}$$

This reaction is sometimes called "solvothermal" synthesis when the solvent is nonaqueous, contrary to hydrothermal, where the solvent is always water (Demazeau 2010). Both types of synthesis refer to similar chemical processes. Cundy and Cox (2003) conceptualized the process as a reactive process of multiphase crystallization involving amorphous and crystalline solid phases in at least one liquid phase.

© Springer Nature Switzerland AG 2019
R. Chaves Lima et al., *Environmentally Friendly Zeolites*, Engineering Materials,
https://doi.org/10.1007/978-3-030-19970-8_2

Another form of synthesis uses of ionic liquids, whose main characteristic is a melting point lower than 100 °C (Dupont et al. 2000; Parnham and Morris 2007). In this case, the ionic liquids are typically organic compounds that have dual roles as solvent and structure-director (Meng and Xiao 2013; Vinaches et al. 2017). This type of synthesis is known as ionothermal and is considered to be green chemistry.

2.2 Physical Chemistry of Zeolite Crystallization

Controlling attributes such as size, habit and degree of crystallinity is imperative for production of adequate crystalline materials. The characteristics of the material depend on the way the reactive species behaves during the synthesis. Although the formation of nanoporous materials may be more complex than the crystallization of nonporous materials (Cubillas and Anderson 2010), the physicochemical principles of nucleation theory and crystal growth are similar and serve as a starting point for understanding certain aspects of nanoporous material nuclei formation (Thompson 2001; Subotić and Bronić 2003).

Zeolites are thermodynamically metastable materials whose synthesis finishes after a long time spent as a denser phase, usually quartz. Thus, zeolite formation occurs in semiequilibrium. These reactions are controlled by thermodynamic factors and kinetic barriers that act together (Navrotsky et al. 2009; Maldonado et al. 2013). The more open the zeolitic structure is, the more metastable the zeolite is, and the favored reaction direction transforms the zeolite into denser structures following the Ostwald rule of successive transformations (Cundy and Cox 2005; Maldonado et al. 2013).

Examining the context concerning the chemical equilibrium of the hydrothermal crystallization reaction of saturated solutions, as in any chemical reaction, the Gibbs free energy (ΔG) varies during the process. When referring to zeolite synthesis, this variation occurs as a function of temperature, pressure and number of moles of the reactive species: $G = f(T, P, n_1 \ldots n_i)$ (Byrappa and Yoshimura 2013). The Gibbs free energy (Eq. 2.2) can take the form of the partial differential equation for nonequilibrium systems (Levine 2009), for example, in zeolitizations (Eq. 2.2) as follows:

$$dG = \left(\frac{\partial G}{\partial T}\right)_{P,n} dT + \left(\frac{\partial G}{\partial P}\right)_{T,n} dP + \sum_{i=1}^{k} \left(\frac{\partial G}{\partial n_i}\right)_{P,T,n_j \neq i} dn_i \qquad (2.2)$$

where $n_j \neq i$ means that in the partial derivative, only the number of moles of species i changes while the quantities of the other species in the system are constants. The variation of the free energy with the number of moles of the species (Eq. 2.3) is equivalent to the chemical potential, μ (Levine 2009) as follows:

$$\left(\frac{\partial G}{\partial n_i}\right)_{P,T,n_j \neq i} dn_i = \mu_i dn_i \tag{2.3}$$

If the pressure and the temperature are assumed to be constant during the process, the postintegration equation for all the species involved in the reaction is Eq. (2.4) (Levine 2009; Byrappa and Yoshimura 2013) as follows:

$$G_{P,T} = \sum_{i=1}^{k} \mu_i n_i \tag{2.4}$$

The processes that intervene in the crystallization are generally compiled in a single event known as the induction period, characterized by the time interval between the establishment of supersaturation and the appearance of crystals that are detectable by X-ray diffraction (Cubillas and Anderson 2010).

The induction period, τ (Eq. 2.5), involves substeps such as the relaxation time, t_r, or equilibrium reactions occurring in the synthesis gel in the temperature range, the time t_n required for nucleation and the time t_g required for the nuclei to reach a detectable crystal size (Cundy and Cox 2005) as follows:

$$\tau = t_r + t_n + t_g \tag{2.5}$$

The formation of crystals from solution begins with the nucleation process, which is of great importance in the synthesis of zeolites since the nucleation process determines the type of structure formed (Pérez-Pariente 1995). Cubillas and Anderson (2010) describe nucleation as a series of atomic and molecular processes by which the reactant species rearrange to form critical product cores large enough to grow irreversibly. There are two types of nucleation: primary nucleation, where the formation of nuclei is induced by the solution itself (homogeneous) or by stagnant particles in the system (heterogeneous) and secondary nucleation, where crystals that have already formed seed the solution, which is a special case of heterogeneous nucleation (Thompson 2001; Cundy and Cox 2005).

In energy terms, an important condition for nucleation is the nonequilibrium of the system, achieved by supersaturation of the reaction mixture. Supersaturation is defined as the difference between the chemical potentials of the species in solution, μ_s, and the bulk of the crystal, μ_c (Eq. 2.6) as follows:

$$\Delta\mu = \mu_s - \mu_c \tag{2.6}$$

The chemical potential can be obtained in terms of the standard chemical potential, μ_0 (Eq. 2.7) as follows:

$$\mu = \mu_0 + kT \ln a \tag{2.7}$$

This expression produces Eq. (2.8) for the variation of the chemical potential (Mullin 2001; Cubillas and Anderson 2010) as follows:

$$\Delta\mu = k_B T \ln S \tag{2.8}$$

where k is the Boltzmann constant, and S is the supersaturation ratio. When the global chemical potential tends towards zero, the product of the zeolite reaction system of tends to be denser.

The entropy S depends on the system and, for solutions, is defined as the ratio between the product of the activities of the molecules in the crystal at a given instant, $a_i(m^{-3})$, and equilibrium, $a_{i,e}(m^{-3})$, both elevated to the number of ions i in the crystal (Eq. 2.9) (Kashchiev and Rosmalen 2003; Cubillas and Anderson 2010) as follows:

$$S = \exp\left(\frac{\Delta\mu}{kT}\right) = \frac{\prod a_i^{n_i}}{\prod a_{i,e}^{n_i}} \tag{2.9}$$

The complex nucleation process can occur when $S > 0$, implying $\Delta\mu > 0$, which indicates that the reaction medium is supersaturated. While this factor remains high, nucleation is spontaneous, and the nucleation rate increases (Kashchiev and Rosmalen 2003).

The total change in free energy is equal to the sum of the variation in free energy between the surface of a particle and its bulk, ΔG_s, and the variation in free energy between a large particle and the solution solute, ΔG_V (Eq. 2.10) as follows:

$$\Delta G_T = \Delta G_s + \Delta G_V \tag{2.10}$$

A critical nucleus or cluster is composed of a number of molecules n within a volume V, so each nucleus adopts a more stable form containing $(4/3)\pi r^3/V$ molecules, assuming a spherical shape. ΔG_s is proportional to r^2, while ΔG_V is proportional to r^3, so Eq. (2.11) takes the following form (Mullin 2001; Cubillas and Anderson 2010) as follows:

$$\Delta G_T = 4\pi r^2 \gamma + \frac{4}{3}\pi r^3 \frac{\Delta\mu}{V} \tag{2.11}$$

where r is the cluster radius, γ is the interfacial tension or surface energy, and $\Delta\mu/V$ is the variation in free energy per volume, ΔG_v. Since the two free energy terms have opposite signs, the nuclei total free energy of formation has a maximum value representing the energy barrier that needs to be surpassed for nucleation to occur, ΔG_c, and a critical radius, r^*, for the nucleus to be entropically favorable in the system (Fig. 2.1) (Mullin 2001; Cubillas and Anderson 2010; Schweizer and Sagis 2014).

The values referring to the critical coordinates can be obtained by simple mathematical operations. The value of r^* can be found by derivation (Eq. 2.12) as follows:

Fig. 2.1 Total free energy as a function of cluster size, where r_c is the size of the nucleus. Reproduced (adapted) with permission from Wu et al. (2016), Copyright (2016) American Chemical Society

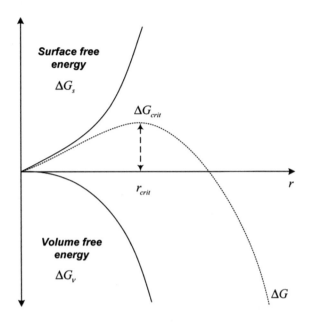

$$\frac{d\Delta G}{dr} = 8\pi r\gamma + 4\pi r^2 \Delta G_v = 0 \rightarrow r^2 = -\frac{8\pi r\gamma}{4\pi \Delta G_v} \rightarrow r^2$$

$$= -\frac{2r\gamma}{\Delta G_v} \rightarrow r_c = -\frac{2\gamma}{\Delta G_v} \tag{2.12}$$

ΔG_c substituted in the general equation results in Eq. (2.13) as follows:

$$\Delta G_c = 4\pi\left(\frac{4\gamma^2}{\Delta G_v^2}\right)\gamma + \frac{4}{3}\pi\left(-\frac{8\gamma^3}{\Delta G_v^3}\right)\Delta G_v$$

$$\Delta G_c = \left(\frac{16\pi\gamma^3}{\Delta G_v^2}\right) + \left(-\frac{32\pi\gamma^3}{3(\Delta G_v^2)}\right)$$

$$\Delta G_c = \frac{16}{3(\Delta G_v)^2}\pi\gamma^3 \tag{2.13}$$

These equations show that the energy barrier for the formation of the nucleus decreases with increasing supersaturation, since this relationship implies an increase in the chemical potential (Cubillas and Anderson 2010). Then, this inference results in Eqs. (2.14 and 2.15) as follows:

$$r_c = -\frac{2\gamma V}{k_B T \ln S} \tag{2.14}$$

$$\Delta G_c = \frac{16\pi\gamma^3 V^2}{3(k_B T \ln S)^2} \tag{2.15}$$

Zeolite synthesis involves at least one liquid phase and amorphous and crystalline solid phases, and it is not easy to measure supersaturation (Cundy and Cox 2005). However, it was possible to theorize the roles of certain factors in the crystallization process.

The kinetic aspect of nucleation is the number of nuclei formed over a given time period (Eq. 2.16). The nucleation rate can be expressed by an equation such as that of Arrhenius (Mullin 2001) as follows:

$$J = A \exp\left(\frac{-\Delta G_c}{kT}\right) \tag{2.16}$$

where A is a factor that depends on supersaturation, and ΔG_c is related to the activation energy for nucleation.

Thus, the nucleation rate is virtually zero until the chemical potential is reached. Consequently, the supersaturation reaches the critical value, the activation energy; subsequently, the nucleation rate grows exponentially until reaching a maximum from which the rate begins to decrease (Mullin 2001; Cubillas and Anderson 2010). This decline occurs due to changes in the viscosity of the reaction medium that affect the mobility of the species and the formation of nuclei (Mullin 2001).

The free energy calculus in this theory cannot be used with precision for porous materials. In this case, the high specific area of the nuclei may imply possible changes in the contact surface between the porous structure and the solution, modifying the solid-liquid interface, which affects the interfacial tension factor γ and results in lower activation energies than those expected by classic theory. Moreover, the reaction kinetics keep the concentration of nuclei low by interfering with the thermodynamic assumptions (Pope 1998).

However, if during nucleation, the solvent molecules and cations behave analogous to how these components behave during the stabilization of already formed crystals, this problem may be avoided. However, only semiquantitative assessments can be made (Pope 1998).

Another way of expressing the activation energy per mole of nucleation in terms of the induction period in a linear relationship with temperature[1] is according to Eq. (2.17) proposed by Zhdanov and Samulevich (Qinhua and Aizhen 1991) as follows:

$$\frac{d \ln \frac{1}{\tau}}{d \frac{1}{T}} = \frac{\Delta G_c}{R} \tag{2.17}$$

After the formation of the first nuclei, the second stage of crystallization begins, that is, crystal growth. Crystal growth occurs by the assimilation of the components in solution and the unstable nuclei. This process takes place through the nucleation of phases as dictated by the composition of the solution, followed by nuclei growth through the incorporation of the content still in solution, and requires orientation

---

[1] See Fig. 4 in Culfaz and Sand (1973).

according to the crystallography of the material in addition to the transport of matter (Randolph and Larson 1988; Thompson 2001).

Thus, nucleation and crystallization occur concomitantly to a certain extent, but the growth stage dominates when the chemical potential is not favorable for nucleation. The crystal growth can be understood as a series of processes involving mass transfer in and through the solution (growth zone) and processes of incorporation and movement of atoms on the surfaces of the crystals themselves (Cubillas and Anderson 2010). Byrappa and Yoshimura (2013) included recrystallization of the solids, including dissolution in the liquid phase, prior to mass transfer.

Although the process of reagent transport in solution to the surface of the stable nuclei is essentially diffusion (Randolph and Larson 1988), a study carried out with the LTA structure (Bosnar and Subotić 2002) concluded that the growth of zeolitic crystals is controlled via interfacial processes. This hypothesis is supported by the discrepancy between zeolitization activation energies that, according to Cubillas et al. (2013), are between 45 and 90 kJ mol^{-1}—for example, the zeolite Y (FAU) with $\Delta G_c = 71.128$ kJ mol^{-1} (Qinhua and Aizhen 1991)—and those values expected for diffusion mechanisms (12–17 kJ mol^{-1}).

According to Bosnar and Subotić (1999), experimental evidence indicates that the growth of zeolite crystals depends on the silicon and aluminum concentrations in the liquid phase of the system and to demonstrate this, an equation was proposed (Eq. 2.18) as follows:

$$\frac{dL_m}{dt_c} = k_g f(c) = k_g f\left(c_{Al}, c_{Al}^*, c_{Si}, c_{Si}^*\right)^r \qquad (2.18)$$

where $\frac{dL}{dt_c}$ is the rate of change of the crystal size over time, k_g is a linear crystal growth constant, $f(c)$ is the concentration factor, c_x is the concentrations of silicon and aluminum in the liquid phase during crystallization, c_x^* is the silicon and aluminum concentrations corresponding to the solubility of the zeolite under certain synthetic conditions, and r is the Si/Al ratio of the zeolite formed.

The results obtained with this theory (Fig. 2.2) agree with the kinetic model proposed by Davies and Jones for the growth and dissolution of crystals. According to this model, the formation of a monolayer of hydrated ions with constant composition occurs at the growing crystal-solution interface. In this way, the growth rate is controlled by the flow of species at the solution-crystal interface (Davies and Nancollas 1955).

Many scientists have suggested mathematical and computational means of describing the crystal growth process, relating physical and even geometric factors to the free energy fractions (Mullin 2001; Leite and Ribeiro 2012). However, the high complexity of the process makes it difficult to establish a true mechanism, even with the appearance of in situ techniques, such as atomic force microscopy (AFM) (Cubillas et al. 2013; Lupulescu and Rimer 2014) and small angle X-ray scattering (SAXS) (Sankar et al. 2007).

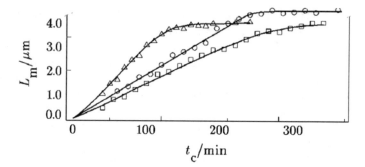

Fig. 2.2 Graph of the size of zeolite A crystals as a function of time. The triangles, circles and squares represent experimental measurements, and the curves were calculated by applying Eq. (2.17). Reproduced with permission from Bosnar and Subotić (2002), Copyright (2002) Croatian Chemical Society

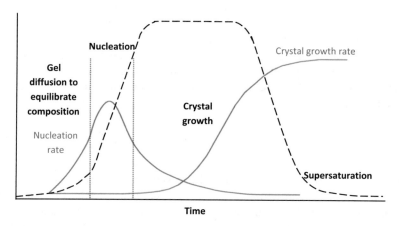

Fig. 2.3 Schematic representation of the nucleation and crystal growth processes and the supersaturation of the system in zeolite synthesis

In practice, the nucleation and growth processes of zeolite crystals are hardly distinguishable but may be summarized on a graph (Fig. 2.3), which shows how these processes occur with the reaction time.

To diminish the mistakes surrounding the global process, zeolite crystallization curves generally have a sinusoidal shape (crystal growth rate in Fig. 2.3), first shown in work by Flanigen and Breck in the 1960s, and this format can be correlated with a Kolmogorov equation (Eq. 2.19) (Valtchev and Mintova 2009) as follows:

$$\frac{m_t}{m_f} = X_t = 1 - \exp\left(-B_t^n\right) \tag{2.19}$$

where m_t and m_f are the masses of the crystals in the product at time t and in the final product, respectively, B is a constant applicable to homogenous nucleation, and n is a constant that depends on the conditions of nucleation and crystal growth. Even so, the literature presents S-shaped curves obtained in ways that do not use the masses of products.

The hydrothermal synthesis of a zeolitic aluminosilicate includes the breakdown of the Si–O and Al–O chemical bonds of the precursors, often to an amorphous structure that rearranges, forming the Si–O–Al bonds of the crystalline structure. Due to this step, the enthalpy variations must be modest (Petrovic et al. 1993; Piccione et al. 2002).

In reality, the global free energy variation for zeolite synthesis is very small (Cundy and Cox 2005); since $\Delta G = \Delta H - T \Delta S$, the enthalpy is a very significant portion of the energy in the isothermal transition of the amorphous phases to crystalline phases.

Several measurements and modeling for enthalpy calculations of zeolites and zeotypes have been reported in the literature (Petrovic et al. 1993; Piccione et al. 2002). Many of these calculations have been performed using calorimeters with hydrofluoric acid and thermodynamic data for quartz, which is one of the more stable phases of silicon dioxide.

Mathieu and Vieillard (2010) proposed a model for estimating the formation enthalpy of hydrated and dehydrated zeolites based on the formation enthalpies of the oxides that form the reaction gel, the crystalline lattice atomic density parameters, and the electronegativity differences. The model obtained results that differed no more than 21.05 kJ mol^{-1} from experimental measurements. The formation enthalpy of a zeolite was concluded to be a function of the density, the Si/Al ratio and the nature of the compensation cations. This model was valid and useful for understanding the behavior of zeolites in saturated solutions and as a function of temperature, among other applications.

Finally, Anderson et al. (2017) proposed a new kinetic model for a three-dimensional partition adapted to porous solid growth based on the Monte Carlo computational method. This model demonstrates the formation of a zeolite from metastable entities on the surfaces of the crystals, the "growth units". These calculations have the potential to estimate the way growth develops and the energies involved in the process (Fig. 2.4).

Although the mechanisms of zeolite synthesis are still being studied, certain approaches have been useful for discussions and have expanded the perspective regarding the formation of these materials. Currently, the size and morphology of zeolite crystals can be controlled to a certain extent by studying the different synthesis parameters.

2.3 Mechanisms of Zeolite Formation

Given that zeolite synthesis involves complex events, many researchers have proposed mechanisms for crystal formation even though the way the precursors behave is very complicated and not well understood.

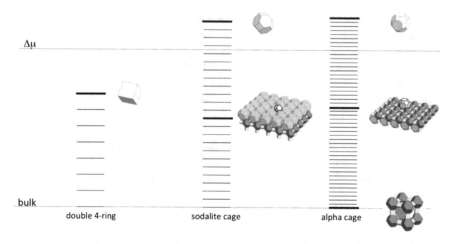

Fig. 2.4 Kinetic three-dimensional partition model applied to the LTA structure. An illustration of the energy levels for the structure and each of its components. Reproduced (adapted) with permission from Anderson et al. (2017), Copyright (2017) American Chemical Society

Despite their complexity, the mechanisms have evolved with the experimental and/or theoretical evidence. According to Grand et al. (2016), there are two different schools that describe the crystallization of zeolites. The classic school describes the crystallization in a way similar to the previous discussion. The nonclassic school comprises the aggregation and annexation processes of entities in the system.

The first attempt to explain how zeolites are formed was reported by Barrer et al. (1959) and has pedagogical characteristics. This study commented on the free energy balance and proposed the idea of SBUs as entities that form unit cells by polymerization reactions (Cundy and Cox 2005).

The ideas proposed by the Barrer group were supported by the Raman spectroscopy tests of McNicol et al. (1972, 1973) in that hydroxylated silicon and aluminum species were detected in both phases. These researchers concluded that the crystallization mechanism occurs in the solid-state by observing the changes.

In 1960, Flanigen and Breck launched a kinetic tracking study with X-ray diffraction synthesis. This study showed that zeolitization consists of an extended period of induction during which the nuclei, which could be SBUs and not necessarily whole unit cells, reach the critical size via heterogeneous nucleation followed by rapid crystal growth (Cundy and Cox 2005).

The growth of the crystals occurs by polymerization/depolymerization processes of the Si–O and Al–O bonds, catalyzed by the basic media (Cundy and Cox 2005), as represented by Eqs. (2.20 and 2.21) as follows:

$$T{-}OH + {}^{-}O{-}T \rightleftarrows T{-}O{-}T + OH^{-} \tag{2.20}$$

$$T{-}OH + HO{-}T \rightleftarrows T{-}O{-}T + H_2O \tag{2.21}$$

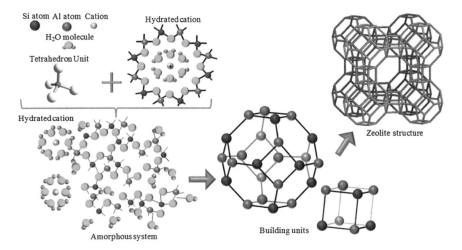

Fig. 2.5 Schematic of Flanigen and Breck's mechanism

These reactions predominantly involve the solid phase (Fig. 2.5) but have some influence on the liquid phase, through which TO_4 units form the subsequent arrangement of SBUs (Cundy and Cox 2005). Shortly thereafter, Breck (1964) added that the polymerization is caused by the hydrated cations of the alkaline solution and occurs by bonding the tetrahedra to form the SBUs.

Subsequently, George Kerr published the results of experiments (Kerr 1966) that led to the chemical reaction in Eq. (2.22) as follows:

$$amorphous\ solids \xrightarrow{fast\ step} soluble\ species + nucleus \xrightarrow{slow\ step} zeolite \qquad (2.22)$$

Kerr argued that a zeolite is formed by dissolving the amorphous gel species and then rapidly depositing the species into the first forming nuclei, which is the rate-limiting step of the reaction and that the direct transformation of all the amorphous content into crystals does not occur.

Kerr assumed that the rate of zeolite formation is a function of the nuclei concentration Z (with first order kinetics) in solution and the solubility S of their precursor species (Eq. 2.23) as follows:

$$\frac{dZ}{dt} = k[S] \qquad (2.23)$$

Kerr proposed that this assumption explained the propositions of Flanigen and Breck.

Additionally, in the 1960s, Ciric (1968) published a detailed article related to the kinetics of zeolite synthesis regarding the amount of adsorbed water. The species that act as the building blocks of the crystalline structure have been postulated to be dimers and tetramers with negative bivalence.

Ciric also reported that saturation equilibrium is maintained by the gel supplying the content in solution, which is consumed by the growth of the crystals until the gel is depleted, and then, everything in the solution is consumed rapidly. Ciric tested the agreement of an equation based on mass transfer and the kinetic approximations from experimental data.

In the 1970s, a paper published by Zhdanov (1974) provided important advances in the understanding of zeolite crystallization. The article deals with the structure of hydrogel—a heterogeneous colloidal system with different compositions in the solid (gel-skeleton) and liquid (solution) phases—and the compositions of the solid and liquid phases as well as the degree of influence of each phase in nucleation and zeolitic crystal growth.

The variations in the atomic ratios of both phases depend on the complex mode of the zeolite synthesis gel composition. Zhdanov argues that the disparity in the compositions of the phases can be explained by polycondensation in the formation of the solid phase from the tetrahydroxyaluminate $Al(OH)_4^-$ and silicate $(OH)_{4-m}Si(O^-)_m$ ions (where m is the $NaOH/SiO_2$ ratio) with different degrees of hydroxylation.

Regarding the structure of the solid phase, evidence indicated the formation of units similar to those found in crystals with tetracoordinated Al atoms and alkali cations compensating the load generated by the $[AlO_4]^{5-}$ units. The degree of hydroxylation tends to increase with decreasing $NaOH/SiO_2$ ratio and dilution, as follows (Eq. 2.24):

$$Si(OH)_4 + m\left(Na^+ + OH^-\right) \rightleftarrows (OH)_{4-m}Si\left(O^-Na^+\right)_m + mH_2O \qquad (2.24)$$

This equilibrium is evidenced on the empirical basis that the Si/Al ratio increases with decreasing gel alkalinity and increasing water content of the gel. This observation showed that the formation mechanisms of the framework blocks in the gel are similar to those occurring during the formation of zeolite crystals.

The liquid phase is bonded to the solid by a semiequilibrium solubility and existence under conditions according to the nature of the silicon and aluminum sources and the gel preparation method (Zhdanov 1974; Cundy and Cox 2005).

At the end of the crystallization, another balance is established between the remaining solution, called the mother liquor, and the crystals. Zhdanov had previously reported that the concentration of aluminate ions in the mother liquor is always lower than that in the liquid phase of the reaction system, while the concentration of silicate ions may be higher or lower.

In the mechanism proposed by Zhdanov (Fig. 2.6), both phases of the gel directly participate in the formation of crystals. The nuclei are initially formed in the liquid phase or interface region between the two phases and grow from different hydrated aluminosilicate ions present in the solution whose structures are associated with SBUs. The gel skeleton, although essentially amorphous, contains units similar to D#Rs that contain aluminol, Al–OH, silanol, Si–OH and Si–O⁻R⁺ groups and are not used in polycondensation reactions due to high reactivity.

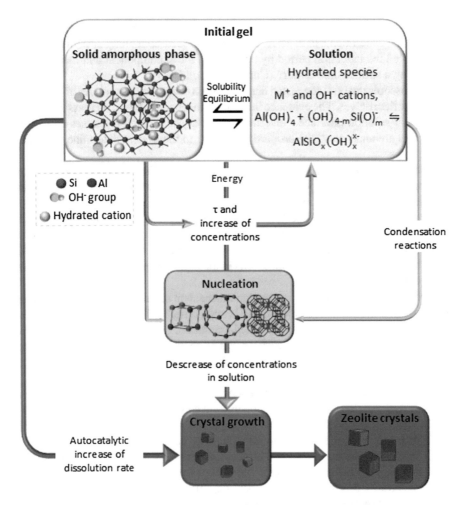

Fig. 2.6 Schematic representation of Zhdanov mechanism

Thus, the probability of condensation reactions increases, through which the aluminosilicate blocks and the clusters are formed. As the nuclei grow, more material must be incorporated into the faces of the crystals, and the rate of dissolution increases; this process was designated "autocatalytic".

The solubility product of the amorphous phase depends on the composition and the temperature; thus, as the temperature increases, more of the solid phase migrates to the liquid phase, and the equilibrium moves to the right.

The reactions continue until all the solution is consumed by the formation and growth of the crystals, and since the zeolites have low solubility, equilibrium is reached with a high percentage of conversion.

Culfaz and Sand (1973) studied the mechanism of crystallization by kinetic monitoring. The experiments of these researchers with seed-containing systems and their use reduced the induction period by eliminating the t_n component and making self-generating a surface unnecessary (Cundy and Cox 2005).

Microscopic analysis suggested that nucleation occurs on the surfaces of seeds that also grow themselves. The evidence indicated a mechanism based on mass transfer to the crystal-solution interface. Nevertheless, the crystallization processes of zeolitic structures are analogous; the properties peculiar to the system determine the speed of the steps.

Aiello et al. (1974) found that the formation of zeolite crystals is preceded by the formation of an amorphous lamellar intermediate that subsequently disappears in heterogeneous nucleation.

Angell and Flank (1977) performed a kinetic study of the solid and liquid phases of a reaction system based on several characterization techniques. These researchers found that before the heat treatment, the system tends to be dominated by silicate ions in the solid phase and aluminate ions in the liquid phase, possibly forming ionic pairs with the sodium cations.

Brunner (1992) proposed a nucleation mechanism called "can and cement". This mechanism assumed the veracity of SBUs and considered the filling of voids in the forming nuclei by water clusters, hydrated ions and/or polyhedra and the electrostatic repulsions. In this mechanism, each crystal is composed of "can" (the voids) and "cement" (the silicate lattice) and forms at once with all the "building materials" assembled.

Gonthier et al. (1993) stated that the empirical evidence did not support the idea that nucleation occurs until the components are consumed as proposed by Zhdanov (1974) and adopted a theory, based on population balance calculations, in which nucleation ceases as soon as all the solid phase is dissolved.

In this work, the existence of real nuclei is questioned, and it is proposed that there is a nucleation directing agent, which could even be small nuclei and that the nuclei arbitrarily originate in regions near the surface of the amorphous phase.

Since the direct extraction of information on zeolite nucleation is difficult (Cundy and Cox 2005), the system behavior was observed during an aging period, and small variants of these concentrations were seen in the assays, suggesting a semiequilibrium solubility between the phases.

Variations in species concentrations between the phases and the mean particle size become large at the beginning of the heat treatment. The results indicated that, after primary nucleation, the conversion of the initial gel to an intermediate amorphous sodium aluminosilicate occurs by mass transfer, which in turn transforms to a crystalline phase by dissolution of the species in secondary nucleation and growth processes of the nucleated zeolitic crystals.

Thus, although the results agree with the growth mediated by the solution, the close contact of the two phases (hydrogel) makes the direct rearrangement of the solid content in the zeolite possible in certain cases since the spaces covered by the chemical species in transport are small.

Boris Subotić and colaborators published studies on the kinetics of crystallization and the alleged autocatalytic nature of nucleation using the nucleation mechanism by solution (Subotić 1988; Subotić and Graovac 1985). In contrast, Eric Derouane et al. (1981) published work where the zeolites were formed either by solution mediation or by direct transformation of the solid.

According to Francis and O'Hare (1998), the mechanisms of solid phase or hydrogel transformation and liquid phase mediated transport are extreme, and any actual process may have similarities to either or both.

Giannetto et al. (2000) presented evidence that the growth rate of the crystals is a function of the concentrations of silicon and aluminum species in solution, that the viscosity of the liquid phase changes the kinetics of the process, that supersaturation is required in relation to product solubility for zeolitization to occur and that the possibility of crystallization without a solid phase suggests a solution-centered mechanism.

Cundy and Cox (2005) systematized the most likely mechanistic ideas up to the date of their revision. At the start of mixing all the reactants, the primary amorphous phase is often formed as a gel or is invisible in the case of synthesis from clear solutions, forming a colloidal phase with similar behavior. This phase is probably heterogeneous, containing amorphous aluminosilicate precipitates, silica and alumina in addition to neutral reagents.

Either by aging or heat treatment, the primary phase is modified by the conversions of the chemical equilibria into the secondary amorphous phase, a pseudosteady-state intermediate that concomitantly establishes the equilibrium between the solid and liquid phases and a characteristic distribution of the silicates and aluminosilicates (Fig. 2.7). This secondary phase has some arrangement but not the periodic structure of a zeolite.

These propositions retain similarities both with the solution-mediated process and the formation of a secondary gel from which crystallite nucleation occurs (relaxation time, t_r). Between the primary and secondary phases, the equilibrium reactions occur. Breck's model reactions are taken as key processes, since they represent the balance between the solid phase and the solution.

Only in the nucleation stage does the structural arrangement inherent to zeolitic structures occur. Aspects such as radius and critical activation energy are used here. For Cundy and Cox, the zeolitic nuclei are usually formed by primary, heterogeneous nucleation and centered in the amorphous phase of the reaction mixture, whether the mixture is colloidal or not.

The spontaneity of the transformation of the secondary phase into zeolite crystals can be estimated as a function of the solubility products of both phases (Eq. 2.25) as follows:

$$\Delta G = -RT \ln\left(\frac{K_{s(gel)}}{K_{s(zeólita)}}\right) \tag{2.25}$$

The increase in order occurs, in part, through the condensation reactions between the Si, Al and O atoms that are catalyzed by OH^- ions successively forming and

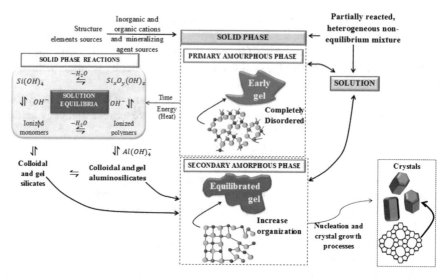

Fig. 2.7 Schematic of the processes in a zeolitization system according to Cundy and Cox

breaking bonds, according to the Flanigen and Breck model. At the same time, the compensating cations form coordinating spheres of oxygenated species with favorable energies and geometries, minimizing the potential energies of molecular arrangements and therefore structuring the zeolitic units.

The hydrated cations act as mediators of the structural ordering from the condensation reactions. Intermolecular forces act by changing the spheres of hydration of the silicate and aluminosilicate units present in solution (coordinating centers) in geometries favorable to the generation of localized regular periodic structural units.

Eventually, the cation may have enough of its hydration sphere replaced by the network of donor oxygen to remain in the site created or to migrate to another region where the cation is incorporated into the network in a similar way due to the balance of charges of the reaction system.

In the meantime, regions randomly distributed by the secondary phase with a higher degree of organization (protonuclei) are formed that may eventually grow or be dissolved by incorporating sections that form the critical radius clusters. The growth occurs from the propagation of very periodic sections, forming a topological rearrangement and resembling an isomerization.

With respect to the equilibria, the reactions in the inverse direction of condensation, which are characteristics of the component t_r of the induction time, become less spontaneous and are overcome by the framework growth by the component t_g. These events do not imply the extinction of the "inverse" species, nor that the arrangement formed is already perfectly crystalline. Therefore, nucleation is a phase transition whereby a system with a certain degree of periodicity becomes a structure ordered enough to form centers of growth for the crystalline lattice.

A somewhat tangent perspective to the chemical context is that nucleation can be thought in terms of the system dynamics, where the constituent fragments collide and are separated, and the flow of species is an essential factor.

In the course of crystallization, there is a point at which nucleation stops, and the most commonly observed step in the synthesis of zeolites, crystal growth, predominates. The growth stage is usually studied kinetically.

Measurements show that, although the rate distribution is broad, the growth rates (in $\mu m\ h^{-1}$) of the zeolites are considerably lower than those for simple ionic salts or molecular compounds; these data corroborate the idea of polymeric piece-by-piece assembly with silicate and aluminosilicate species. The growth rate of the zeolite crystals is linked to the kinetics of the equilibrium reactions of formation and the cleavage of T–O–T bonds.

Graphs of crystal size *versus* time reveal steady linear growth until components start to run out. Thus, a mechanism controlled by the incorporation of matter at the crystal surfaces is hypothesized, because if the process was dependent on the dissolution, the graphs would be curved due to the increasing reagent flow required, but the reagents approach extinction, and the energies are too high to facilitate a diffusional process.

The predominant mode of incorporation could be the adsorption of layers, such as in the Kossel model (Cundy and Cox 2005). The incorporation mechanism describes growth units with undefined specific compositions that are probably monomeric and lose a certain degree of freedom when approaching the surface of the critical nuclei. These growth units are then adsorbed on the surface of the crystal and migrate to a site of high energy; the number of bonds is maximized until the layer is complete, and no further growth can occur without creating another crystallization center by two-dimensional nucleation or a "core island monolayer". The various interfacial defects imply sites of different energies.

This mechanism can represent the process in crystals with conventional sizes (>0.5 μm); for colloidal and nanometric crystals, aggregation processes can significantly contribute. The aggregate growth model of Dokter et al. (1995) (Fig. 2.8) uses the idea that particles with diameters of 3–4 nm aggregate and densify, generating 6–7 nm nuclei that are amorphous to X-ray diffraction (XRD) and aggregate with each other, which generates visible particles of approximately 50 nm.

Forms of aggregation somewhere between amorphous and partially/totally crystalline particles are proposed, including the direct and inverse reactions of condensation and dissolution/incorporation. Either way, there is consensus regarding that the foundation for growth is the increment of matter that is organized, establishing the building units at the growth sites.

Cundy and Cox (2005) stated that there was a preference for the reactions involving the solution because the activation energies are smaller, to the detriment of the solid transformations. Although solid transformations are possible, the transformation may be limited to the case where a growth unit is carried from one region to other adjacent regions instead of being released in solution and transferred by ionic mobility to a growing crystal, similar to the Angell and Flank propositions.

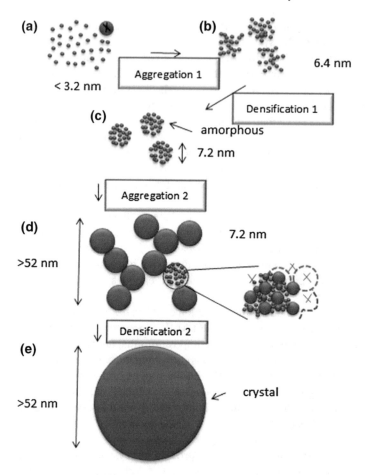

Fig. 2.8 Model of growth by aggregation. Reproduced (adapted) with permission from Dokter et al. (1995), Copyright (1995) American Chemical Society

The complexity of multicomponent hydrothermal reactions is high, and no final conclusions can currently be drawn. In many cases, the processes described in a relatively didactic way can occur simultaneously. For Francis and O'Hare (1998), the major obstacle in understanding the synthesis of molecular sieves is the absence of a universal mechanistic model.

Each structure and each synthesis condition or raw material source can take different routes for the formation of the final product. Nevertheless, the considerations mentioned so far may help in understanding the hydrothermal synthesis process (Byrappa and Yoshimura 2013), and some works propose "general" candidate mechanisms for synthesis.

The generalized mechanism proposed by Cundy and Cox (2005) (Fig. 2.9) is supported by the ideas previously discussed.

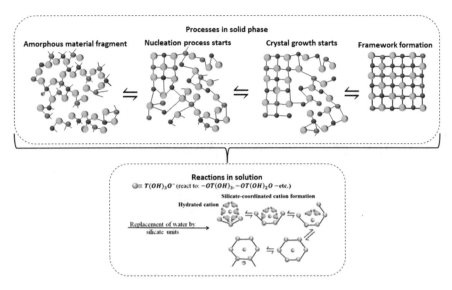

Fig. 2.9 Illustration of Cundy and Cox's mechanism. The processes occurring in the solid phase to construct the 3D framework occur by the mediation of the hydrated cation organization of the species equilibrium reactions

However, Do et al. (2014) proposed a synergistic mechanism (Fig. 2.10) where the crystals form both in solution and by solid-state transformation and combine to produce the crystals of the final product (monoliths).

Crystallization is initiated by the action of Na^+ and OH^- ions rapidly dissolving the gel, generating the soluble species T–O$^-$ and T–OH from T–O. These species react with the hydrated cations, forming SDA–T–O species to aggregate primary subcolloidal particles that then aggregate and densify into an equilibrium gel composed of surface condensed aggregates in the matrix of the primary gel. Nucleation occurs in this gel in equilibrium by reorganization and can move nuclei through the solid-liquid interface and the solution.

These aggregation processes can be represented by the model of Dokter et al. (1995) but occur both in the precipitated amorphous phase of the aluminosilicate solution and in the solution and its interface with the solid phases.

The solid-state crystallization occurs in a type of amorphous tertiary phase that originates from transformations of the primary phase, as in the model of Ren et al. (2012) for heterogeneous systems (Fig. 2.11). The nucleation that occurs in the solution and at the interface better agrees with the catalytic autonucleation model, because the process happens in an intermediate phase of the gel.

The subcritical particles self-assemble immediately via electrostatic attraction between the positively charged species and the TO$^-$ groups on the surfaces of the nuclei into crystals that can grow in crystalline aggregates by nucleation. These attachments also generate aggregates of aluminosilicate anions due to TO$^-$ groups that do not react and whose negative charges are continuously neutralized. While

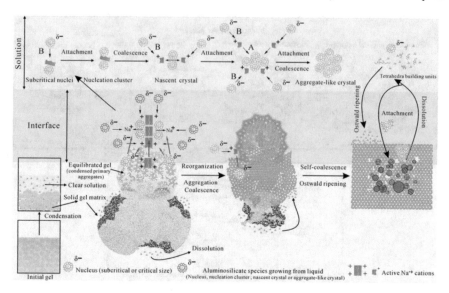

Fig. 2.10 Synergic mechanism of Do et al. Reproduced (adapted) with permission from Do et al. (2014), Copyright (2014) American Chemical Society

growing, the M$^+$ cations are expelled from the T–O–T network and immediately react with the negatively charged aggregates from the solid and liquid phases.

In the final steps, the self-recrystallization via supersaturation processes form crystals by dissolving the predominant less stable crystals. The [OH$^-$] in solution increases with the crystallization time by accelerating the formation of soluble species, increasing the speed of crystallization and causing the crystals to detach from those growing in the gel matrix. The formation of anionic aggregates (semiordered precursor) may be due to the condensation of subcritical nuclei with dissolved species. Apparently, the mineralizing ions released from the agglutination process also rupture the less stable T–O bonds of the crystalline particles.

The studies on the processes of the formation of zeolite crystals continue, and the use high-resolution characterization techniques combined with the previous findings are especially becoming points of reference.

Grand et al. (2016) summarized the proposal for hydrogel crystallizations and colloidal suspensions and concluded that the interactions of diverse forces between the building units occur to form the covalent network of zeolites, that the increase in [NaOH] during the initial polymerization results in the formation of randomly aggregated small open structure particles, and that the composition of the gel particles approaches the stoichiometric composition of the zeolite as the equilibrium between the phases of the system is reached.

The most important points to date are: precursor species are formed in the mixture and their sizes depend on the synthesis variables, the dynamic processes of aggregation and densification create amorphous species crucial for nuclei to appear,

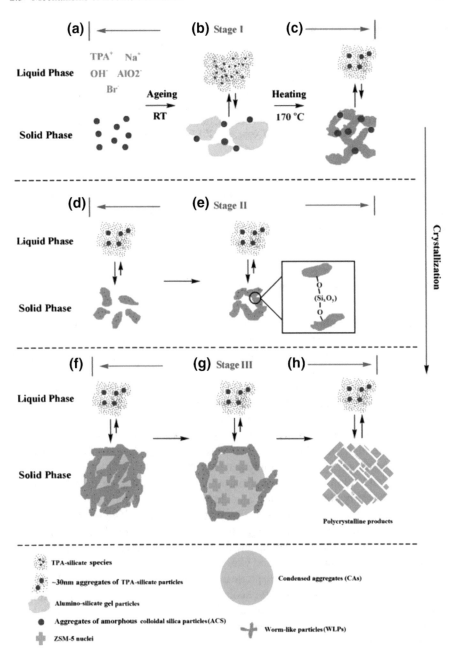

◄**Fig. 2.11** Model of Ren et al. (2012) for zeolite crystallization. The aluminate ions in solution react with the silica particles in (**a**) and (**b**), which are dissolved, forming Si–O–Al bonds on the particle surfaces. During heating, the condensation reactions form earthworm particles (WLPs) (**c**), which are then separated by disintegration and dissolution of the amorphous particles (**d**). The WLPs aggregate by the condensation of silanol groups, whose concentrations increase on the WLP surfaces (**e**) and then densify by coalescence (**f**). The high concentrations of the system species in the amorphous matrix of the condensed aggregates promotes the formation of the zeolitic nuclei (**g**) until the content is consumed in the nucleation and growth of the crystals (**h**). Reproduced (adapted) with permission from Ren et al. (2012), Copyright (2012) American Chemical Society

nucleation occurs in amorphous particles, and crystal growth occurs in the amorphous particle interior followed by the addition of nanoparticles. The only differences between the zeolites obtained by dense gels and those by colloidal suspensions are the morphology and particle size (Grand et al. 2016).

2.4 Influence of Synthesis Parameters

Since zeolites are traditionally synthesized from saturated solutions with defined compositions, the temperature and pressure must be determined. Variations of any of these parameters allow syntheses of materials with diverse structures and chemical compositions (Giannetto et al. 2000).

Although there are indications that the overall process of zeolite crystallization is managed kinetically, a range of thermodynamic variables (temperature, pressure and chemical composition) and the handling of the reaction system before and during the synthesis affect the substeps involved, the spontaneity of the phases formed at any given moment in the reactions, and the intrinsic details of the crystals such as their dimensions and morphologies (Cundy and Cox 2005).

The Ostwald rule of successive transformations states that an unstable system does not necessarily directly produce a stable phase, and less stable phases may arise before reaching the most stable phase by dissolution and self-reorganization with slight variations in free energy. The phases maintain the material balance between them by a type of disproportionation, and the transformations can be complete or not depending on the system (Cundy and Cox 2005).

Another phenomenon inherent to phase transitions is Ostwald ripening, adopted by Van Santen et al. (1986) as a model for nucleation, which relates the solubility product to crystal growth in the sense that particles formed in a supersaturated solution tend towards minimal surface energy. The phenomenon is usually significant when the system is filled with particles with radii ≤ 1 μm (Cundy and Cox 2005).

This phenomenon results in a critical cluster size from which the larger nuclei grow and the smaller nuclei resolve, and both the ratio between the solubilities of the two nuclei groups and the critical radii emerge from the aforementioned expressions. Cundy and Cox (2005) argue that there is not much evidence to substantiate the importance of these terms in zeolite synthesis since the stability of a zeolite in its parent water depends heavily on its purity.

In the most common cases of zeolite synthesis, where the reaction system has the generic composition Na_2O–SiO_2–Al_2O_3–H_2O (Šefčík and McCormick 1999), different zeolite phases, chemical or structural precursors and/or intermediates may be momentarily more stable at a given time of reaction. The target zeolite phases are often equally metastable, and studies are needed to define the optimized conditions for their synthesis (Cundy and Cox 2005).

2.4.1 Temperature

Geological considerations have suggested that the crystallization temperature of zeolites reaches a maximum limit of 350 °C, although certain zeolites crystallize at higher temperatures. The current temperature ranges are: ambient (25–60 °C), low (90–120 °C), moderate (120–200 °C) and high ($\geq$250 °C) (Szostak 1989).

The temperature domain becomes clearly important for the control of zeolite synthesis by thermodynamic considerations regarding the role of energy in the formation and growth of nuclei, where temperature increases facilitate overcoming the activation energy for the formation of nuclei.

Zhdanov (1974) studied the effect that thermal energy has on the composition of the reaction mixture. He compared data obtained at room temperature gel composition data after 4 h of heat treatment at 90 °C and reported an increase in the concentrations of all components of the blend in the liquid phase, which is expected considering that solubility usually increases with increasing temperature in endothermic processes. Therefore, the increase of the species concentration in the solution is part of the crystallization process.

The more energy that is injected into the system, the shorter the nucleation stage, and the more the growth of the crystals is favored. The kinetic relationships between heat, nucleation and growth were evident in the discussion of crystal chemistry. For example, Freund (1976) noted that nucleation works best at room temperature because crystal growth does not impair nucleation at this temperature.

The temperature also influences the kinetics of the synthesis due to the tendency of increasing temperatures to decrease viscosity and the dielectric constant of water, facilitating mobility (Yu 2007) and increasing the solubility of the species (Szostak 1989) and the condensation equilibria (Xu et al. 2007). However, it is not possible to determine the amount of zeolite formed.

Thus, the temperature can change the induction time and the thermodynamic stability of the zeolitic phases obtained. Higher temperatures tend to result in smaller pores in greater numbers due to the escape of water, which stabilizes the pores. There is a dependence between time and temperature, so one value can increase or decrease depending on the other (Szostak 1989).

According to Cundy and Cox (2005), an increase in temperature may cause the synthesis processes to overlap by advancing the reaction kinetics. Therefore, the sizes of the crystals are influenced by the temperature through its effects on the phase transition enthalpy, the activation energy coefficient and the Gibbs-Thomson interfacial energies (Madras and McCoy 2003).

2.4.2   Gel Aging

Aging consists of resting the reaction mixture for a period of time before heat treatment, leading to crystallization. For convenience, an aging period is generally conducted at ambient temperature but may be applied at temperatures higher than that of the crystallization itself. For high temperatures, the overall process consists of two independently administered crystallization steps (Cundy and Cox 2005).

The main effect of aging is to promote a certain amount of "separation" between the nucleation and crystal growth. Aging commonly contributes to the generation of areas with certain degrees of order in the secondary phase, called proto-nuclei (Cundy and Cox 2005). With more proto-nuclei, there is less material for incorporation into growth, which alters the zeolitization kinetics, even though not all the clusters reach critical growth values.

The proto-nuclei formed during aging may be inactive by maturing until the system undergoes an increase in temperature, at which time the growth process becomes more prominent, or the system may be inactive under the same conditions long enough for nucleation to proceed. The conversion of proto-nuclei into clusters with critical radii makes cation-mediated reactions viable for initiating growth (Cundy and Cox 2005).

Caputo et al. (2000) studied the crystallization kinetics of homogeneous systems with and without aging using dynamic scattering of light and verified the absence of precursor phases in aging systems.

Because the growth and the bond formation/breakage rates are related, determining the effects of aging requires subjecting the system to different conditions depending on the crystallization time of the zeolite to be formed (Cundy and Cox 2005).

A theoretical study by Cook and Thompson (1988) showed that moderate periods of aging produce nuclei that remain inactive until the system reaches a certain temperature.

The aging process may favor the nucleation stage (which shows that nucleation occurs without large amounts of energy), making it possible to reduce the size of the crystals (Freund 1976; Cundy and Cox 2005) and decrease the time required for crystallization.

2.4.3 Pressure

In hydrothermal synthesis, the autogenous pressure varies between 0.5 and 2 kbar (Ghobarkar et al. 2003). Pressure is a little studied factor, with some importance in research on natural zeolites and synthesis in the presence of a volatile reagent that can alter the pH and concentration of the dissolved species, which can occur when certain organic compounds are in the system (Gies et al. 1998). This application arises because synthesis at temperatures ≥ 100 °C usually works by autogenous pressure.

The term "autogenous pressure" refers to the saturated vapor pressure of water at the temperature used in the synthesis. For synthesis in the ambient temperature range, the water vapor pressure is small.

2.4.4 Stirring or Static Thermal Treatment

It is common to perform hydrothermal zeolite syntheses in a rotary or static oven. Certain structures can be inhibited or formed depending on which type of oven is used (Giannetto et al. 2000).

The agitation contributes to the homogenization of the heat distribution in the system and can promote microfriction between the hydrogel component species (Cook and Thompson 1988; Cundy and Cox 2005). Freund (1976) posited that these collisions characterize secondary nucleation.

2.4.5 Chemical Composition of the System

According to Corma and Davis (2004), there is an "intelligence" in the complex formation of zeolites that must be contained in the reagents. Therefore, the composition of the reaction system is a primary factor for the synthesis of a specific zeolitic phase.

The general composition of a system for the synthesis of an organic-free type zeolitic aluminosilicate is $Na_2O–SiO_2–Al_2O_3–H_2O$. The structure of the zeolite formed can be discussed in terms of the molar ratios of these components.

A primary aspect is the control of the Si/Al ratio (SAR[2]). For example, zeolite A forms at SAR $= 1$ without the aid of a structural leader, an SAR between 5 and 7.5 yields mordenite, and an SAR ≥ 20 may form ZSM-5. Despite these results, the literature states that it is possible to find a domain of formation where composition variation modifies the chemical composition without changing the structure.

The SAR is linked to the determination of the zeolite phases that are formed in a system (Maldonado et al. 2013, Oleksiak and Rimer 2014) and can modify the properties of certain phases. Structures with low aluminum have better acid resistance and thermal stability, which is desirable for catalytic applications. Zeolites with high

---

[2]Silica to aluminum mole ratio abbreviation.

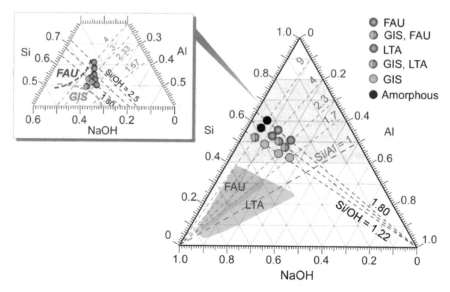

Fig. 2.12 Example of a ternary phase diagram for certain zeolites. Reproduced with permission from Conato et al. (2015), Copyright (2015) Royal Society of Chemistry

SAR have higher cation content, which is interesting for ion exchange applications. There is also a direct relationship between the Si/Al ratio and the temperature such that the higher the silica, the higher the synthesis temperature should be (Szostak 1989).

At a given temperature range, SAR variation may result in different phases, as illustrated in the phase diagrams (Fig. 2.12) of Oleksiak and Rimer (2014), in which several phases were plotted as a function of the molar ratios of Si, Al and MOH.

For example, mordenite is a phase that forms at high SARs in both temperature ranges, since faujasite and gismondite are formed at low SAR.

According to Cundy and Cox (2005), there is a distinction between the behaviors of different reaction systems. Those classic hydrogels with low Si/Al and high alkalinity present practically all the Si and Al in the form of monomers and oligomers as the solid phase in alkaline solution. For high Si/Al and low alkalinity, the formation of silicate and aluminosilicate polymers occurs. A fraction of these entities behave such as species in solution and the rest (colloidal fraction) behave like an amorphous phase invisible to optical detection but equivalent in energy to a visible amorphous solid. These phases cannot be easily separated. Similar behavior occurs in syntheses from clear solutions.

Theoretically, the supersaturation is closely related to the composition of the system and the solubility of the zeolite, although this bonding is not well defined due to the formation of the slightly more soluble precursor phase (Cundy and Cox 2005). Therefore, studies on the equilibrium reactions and the solubility of the component species are of great value.

A starting point is to consider the dissolution of silica (Eq. 2.26) (Lowe 1983) as follows:

$$SiO_2 + 2H_2O \rightleftarrows Si(OH)_{4(sol)} \text{ com } K_s = \left[Si\left(OH_{4(sol)}\right) \right] \qquad (2.26)$$

The silicic acid enters a chain of equilibria in solution, and the more alkaline the medium, the higher the silica solubility limit.

Following theory, the activation energy of crystallization could be estimated as a function of the solubility products of the gel and the zeolite. However, as there are several species in the gel, the equilibrium constants are somewhat laborious to estimate from the data (Cundy and Cox 2005).

2.4.6 Basicity and Acidity of the Synthesis Medium

The pH is one of the most important parameters to control in the synthesis of zeolites. Generally, zeolitizations occur in alkaline or approximately neutral media; the pH usually ranges from 8 to 12. Alkalinity can be treated in the form of the OH^-/Si or H_2O/Na_2O ratios (Lechert 2001).

The pH is an important factor in the synthesis of zeolites together with the thermodynamic quantities. The influence of these factors extends from the concentration of the reactants to the morphology, size and crystallinity of the material through the kinetics, in addition to controlling the supply of OH ions that act as mineralizing agents.

The hydroxide ion is also a complexing agent that combines amphoteric oxides and hydroxides to the solution (Byrappa and Yoshimura 2013). The hydroxide ion content may change the type of structure formed or the composition by exerting control over the solubilities of the Si and Al sources, the degree of polymerization of the silicate species and the polymerization rates of the polysilicate and aluminosilicate ions (Yu 2007).

Low pH values favor the formation of dimeric species and 4-membered rings; for higher values, $Si(OH)_{4-n}O_n^{n-}$ (with n = 1–4)-type monomeric species are more likely (Lechert 2001; Guth and Kessler 1999).

OH^- anions increase the solubility of silica by ionizing the silanol groups and breaking the siloxane bonds (Si–O–Si) (Eqs. 2.27 and 2.28) (Guth and Kessler 1999) as follows:

$$\equiv SiOH + OH^- \rightarrow \equiv SiO^- + H_2O \qquad (2.27)$$

$$\equiv Si{-}O{-}Si + OH^- \rightarrow \equiv SiO^- + HO{-}Si \equiv \qquad (2.28)$$

During crystallization, the second reaction is reversed, causing the SAR of the zeolite to be generally lower than that of the gel (Cundy and Cox 2005).

Due to the increase in the dissolution of the reactants, the increase in the basicity increases the speed and reduces the induction period of the formation of a viable nucleus. The solubility of silica grows almost exponentially with pH and maintains almost all the aluminum in the monomeric form (Byrappa and Yoshimura 2013). Thus, higher pH values favor the formation of aluminum-rich zeolites and lower values favor silicon-rich zeolites (Valtchev and Mintova 2009).

High [OH$^-$] promotes the generation of defects in the network by preventing global binding of the building blocks, and their variation during zeolitization causes the most superficial regions of the crystals to have a higher aluminum content than in their bulk (Valtchev and Mintova 2009).

Fegan and Lowe (1986) found that both nucleation and crystal growth are affected by the alkalinity of the system. The results reported[3] were schematized by Cundy and Cox (2005). These researchers realized that for the system studied, the maximum nucleation was reached at high pH and the inverse was true for maximum growth. Considering the influence of species collision on nucleation, basicity acts to promote the flow of species in solution.

The low pH syntheses have F$^-$ as a mineralizing agent, which acts by increasing the solubility of silica by the formation of SiF_6^{2-} and the aluminate species (Kessler 2001; Lechert 2001), minimizing the occurrence of unbound $\equiv Si-O$ units and serving as template. This effect favors the formation of zeolites with fewer structural defects and low SAR and provides the possibility of synthesis without alkaline cations. Additionally, larger or monocrystalline crystals tends to be obtained due to the decay of the nucleation rate (Yu 2007; Qin et al. 2013).

The use of F$^-$ also decreases the number of transitions between metastable phases, provides structures that are difficult to synthesize in alkaline media, facilitates the incorporation of other atoms and solves the catalytic activation process by directly obtaining ammoniacal forms requiring only calcination (Louis and Kiwi-Minsker 2004).

2.4.7 Sources of Components

The raw materials of zeolite synthesis may exert a great influence on the final result (Zhdanov 1974). Consequently, the different forms the reagents take and the different ways the reagents can be added to the reaction system must be considered in a synthesis study.

(i) Silica sources

Silicon is the primordial structural component of a zeolite, and the silicate species undergo constant modification and transport during a hydrothermal reaction.

The most common silicon precursors are (Yu 2007):

---

[3] See Fig. 30 in Cundy and Cox (2005).

- Fumed silica (pyrogenic): (e.g., Degussa Aerosil®, Cabot Cab–O–Sil®)—fine particles with amorphous structure, a mean particle size of 0.8 μm, specific areas between 50 and 380 $m^2 g^{-1}$ and high purity. The hydrophilic genus absorbs more moisture from the air than the hydrophobic genus (Katz and Mileski 1987).
- Colloidal silica (sols): (e.g., Sigma-Aldrich Ludox®, Merk Klebosol®)—dense, dispersed and stable systems of discrete particles, generally with amorphous structure in water (hydrosol). Their particle sizes are not well defined, although the sizes are in the range 0.1–1 μm (Katz and Mileski 1987; Bergna and Roberts 2006).
- Silica gel: are generally solid porous silicas, are divided into xerogel (dry gel) and aerogel (gel with liquid part replaced by a gas) types, with a mean particle size of 0.8 μm and specific area ranging from 300 to 1000 $m^2 g^{-1}$. The surface is hydroxylated, and the particles are silicate polymers called micelles (Katz and Mileski 1987; Flörke et al. 2005).
- Precipitated silica: alkaline silicate solid (waterglass) or crystal, with hydration water, presenting metasilicate, orthosilicate and disilicate forms depending on the value of n in the formula $Na_2O \cdot nSiO_2$ or Zeosil® silica type (Flörke et al. 2005).
- Silicon esters (e.g., tetraethyl orthosilicate—TEOS)—TEOS or tetraethoxysilane, $Si(OC_2H_5)_4$, the most common member of this category, is a liquid also classified as an alkoxide (silicon bound to organic groups by oxygen) that hydrolyses in contact with water (Howe-Grant 1998).

Freund (1976) noticed that almost all silica dissolve independent of the reaction system preparation method. However, all these sources show different degrees of polymerization and different reactivities for the species formed in solution (Feijen et al. 1994; Yu 2007).

Effects of the use of silicas of diverse natures have been reported. Schwochow and Heinze (1974) found differences in the SiO_2 concentrations in the liquid phase when using amorphous silica and sodium silicate as silica sources, which was attributed to differences in the polymerization of the sources.

Zhdanov (1974) observed variations in the components of the synthesis in the solution of the erionite reaction system, from which this researcher elaborated a plot of the molalities of the components, especially silica, in terms of the reaction time (Fig. 2.13).

According to the continuous curves, the concentrations of the components in solution tend to decrease during the induction period, reaching equilibrium. Only the silica concentration increases before this step and reaches equilibrium late. Abrupt decay of the concentrations of a silica sol system indicates a poor silica solution, and the acute increase of the concentrations during the induction period is due to the dissolution of colloidal particles in the intramicellar alkaline solution. After 48 h of synthesis, the solution reaches the concentration at which crystallization occurs, suggesting no equilibrium with respect to silica for systems containing silica sol.

Derouane et al. (1981) noted that the use of different silica sources influences how crystallization occurs and the material balances in the system phases. Freund (1976) found that the mode of preparation of the silicate solution influenced the

Fig. 2.13 Variations in the concentrations of the components in the liquid phase of the gel obtained from solutions of silicates and aluminates (solid lines) and silica sol (dashed line) during crystallization. Reproduced (adapted) with permission from Zhdanov (1974), Copyright (1974) American Chemical Society

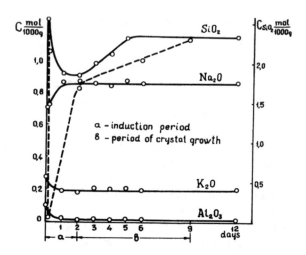

crystallization of the zeolite X. These researchers noted that certain silicas formed the GIS structure rather than the FAU structure.

A study by Hamilton et al. (1993) with 11 silicas of different natures and brands in the NaX zeolite formation showed differences in crystallization kinetics and average crystal size. This study showed the sensitivity of the nucleation with respect to the silica source, mainly due to the disparity in solubilities evident by the turbidity particular to each solution. No differences were detected in the oligomer distributions of the solutions.

Even differences in the silica area may have an effect on the kinetics of zeolitization, because larger areas may favor the emergence and growth of nuclei, according to Cundy and Cox (2005).

Different degrees of polymerization generate a variety of species whose chemical behavior may differ and may modify the final result of the process. Studies with magnetic resonance spectroscopy (^{29}Si–NMR) and thermodynamic propositions have guided the understanding of the nature of these species in an alkaline environment and the physicochemical properties of their solutions.

The solubility of silica is very dependent on the pH of the medium. The condensation rates of monomers decrease as the pH decreases. Thus, when in acidic or near-neutral media, silica is primarily present as monomeric silicic acid, while in alkaline media, these monomers undergo significant deprotonation and condense into oligomers via the formation of siloxane bonds and minimization of silanol groups (Šefčík and McCormick 1997a; Lim et al. 2013).

As for the specific composition of the zeolitic precursor silicate units (Fig. 2.14), spectroscopic analyses by ^{29}Si–NMR, Raman and UV-Raman detect several species containing up to 12 silicon atoms, including cyclic species (Harrison and Hubberstey 1974; Knight 1988; McCormick and Bell 1989; Šefčík and McCormick 1997a; Knight et al. 2006), and even if there are no arguments for considering these species as blocks (e.g., computer studies), these results indicate that simpler species, such

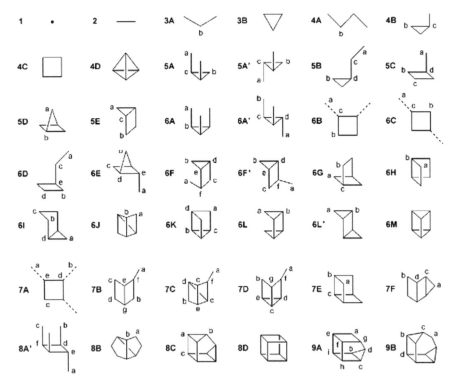

Fig. 2.14 Silicate structures found in a concentrated alkaline solution. The numbers group the structures by the number of Si atoms, the capital letters denote structures in the same group and the lowercase letters denote the chemical shifts of the Si atoms. Reproduced (adapted) with permission from Knight et al. (2007), Copyright (2007) American Chemical Society

as monomers and dimers, are the most likely "bricks" for assembling the crystals (Cundy and Cox 2005).

Šefčík and McCormick (1997a, b, 1999) studied the equilibria of the silica condensation and deprotonation reactions in basic media by means of thermodynamic models. The deprotonation equilibrium exerts reasonable control on the distribution of the silicate species in solution, and the polymerization reactions are credited as an instrument of nuclei formation. Reactions of silicate ions were represented by the following equilibrium expressions (i.e., deprotonation and dimerization reactions) (Eqs. 2.29–2.31) as follows:

$$K_m^i = \frac{\left[\text{SiO}_i(\text{OH})_{4-i}^{i-}\right][\text{H}^+]}{\left[\text{SiO}_{i-1}(\text{OH})_{5-i}^{(i-1)-}\right]} \tag{2.29}$$

$$K_d^i = \frac{\left[\text{Si}_2\text{O}_{i+1}(\text{OH})_{6-i}^{i-}\right][\text{H}^+]}{\left[\text{Si}_2\text{O}_i(\text{OH})_{7-i}^{(i-1)-}\right]} \tag{2.30}$$

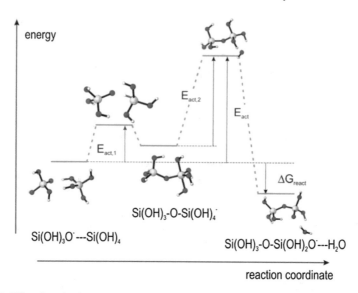

Fig. 2.15 Silica dimerization reaction profile via anionic route mechanism. Reproduced with permission from Van Speybroeck et al. (2015), Copyright (2015) American Chemical Society

$$K_d^{ss} = \frac{[Si_2O(OH)_6]}{[Si(OH)_4]^2} \tag{2.31}$$

where i represents the degree of deprotonation of the monomers and dimers. The importance of the equilibrium constants of these model reactions is justified in the estimation of the solubility constants that are related to the zeolitization activation energy.

The physicochemical preference of zeolitization by monomeric silica species is also suggested in the work of Lindner and Lechert (1994), in which one of the reactions occurring on the surface of a growing nucleus is the condensation of silanol groups (Eq. 2.32) as follows:

$$Zeo \equiv Si-OH + HO-Si \equiv \rightleftarrows Zeo \equiv Si-O-Si \equiv +H_2O \tag{2.32}$$

Computational analysis indicates that polymerization from simpler species is favored (Fig. 2.15). Anions in solution attack the neutral species, forming intermediates that generate dimers by dehydration (Van Speybroeck et al. 2015).

Moreover, the success of the method in predicting the solubility product of zeolite A and crystallization diagrams and the agreement of the results with data from other studies also presented by the authors reinforces the assumption that nucleation strongly depends on monomer entities and does not depend on the SBU-like species detected by NMR.

Experimental results suggest that regarding synthesis with colloidal silica, the system as a whole takes this form by generating an amorphous precursor in colloidal particles, appearing to promote nucleation in solution and restricting the diversity of species. Zhdanov (1974) found that the use of this type of silica implies a low concentration of silicon in solution and that the gel obtained consists of a continuous network of silicon, aluminum and oxygen in which there is no equilibrium between the liquid and solid phases in the reaction system.

For the visible solid phase syntheses typical of pyrogenic silica powder sources, crystallization seems to occur dispersed in the hydrogel matrix, generating greater species diversity (Cundy and Cox 2005; Grand et al. 2016).

Cundy and Cox (2005) commented that the use of pyrolyzed silica sources occurring in the gel matrix is plausible for the zeolitization process, because the concentration gradients remain at the solid-solution interface, and the amorphous solid phase surface provides sites with low free energy for the formation of nuclei, which is equivalent to a reduction in the degrees of freedom of the species. In this sense, the use of silicas with greater surface areas contributes to the acceleration of crystallization by the possible increase in these "nucleation sites".

Grand et al. (2016) indicated that zeolite synthesis with defined properties has to consider the state of the initial reaction mixture (visible gel or clear suspension), because although both contain amorphous particles, their attributes lead to variations, especially in the morphology and size of the crystals.

In the studies by Li et al. (2001) on the aging of the gel, the kinetics were little influenced when TEOS was used, but when using silica sol colloidal (Ludox TM), aging produced more pronounced changes, leading to a decrease of up to 50% in the average size of the crystals. Cundy and Cox (2005) attributed this information to the difference in the reactivities of the silicas, since TEOS systems have high lability, and thus, the depolymerization reactions are less significant.

(ii) Aluminum sources

Aluminum may also come from several sources such as sodium aluminate, gibbsite, pseudoboehmite and even metallic aluminum (Feijen et al. 1994). Aluminum salts can modify the synthesis because these salts have a strong electrolytic effect, so it is advisable to use sources that provide aluminum in the anionic form, such as sodium aluminate (Kühl 2001).

Depending on the aluminum concentration, the solution may contain several polyaluminate species (Gasteiger et al. 1992). However, there is no doubt that the aluminate contributes to the process as monomeric pseudotetrahedral units of $Al(OH)_4^-$, which is the predominant species found in the system assuming the direct nonparticipation of polymeric aluminosilicates (Cundy and Cox 2005; Sipos 2009).

Šefčík and McCormick (1999) adopted the following equilibrium expression for the formation of aluminosilicate ions for their calculations (Eq. 2.33) as follows:

$$K_d^{as} = \frac{\left[AlSiO_k(OH)_{7-i}^{k-}\right]}{\left[Al(OH)_4^-\right]\left[SiO_{i-k}(OH)_{5-k}^{(k-1)-}\right]} \qquad (2.33)$$

where $k = (1, 2, 3, 4)$. The Lindner and Lechert (1994) model, in which the surface of a growing zeolitic crystal has three functional groups: $[Zeo{\equiv}Al{-}OH]^-$, $Zeo{\equiv}Si{-}OH$ and $Zeo{\equiv}Si{-}O^-$, also uses monomeric species and explains the importance of the silicon and aluminum concentrations in the solution for the maintenance of crystal growth.

According Livage (1994), in addition to the condensation of silanol groups, the most likely reactions for the incorporation of entities into clusters include nucleophilic attack on aluminum centers (olation and oxolation reactions) (Eqs. 2.34 and 2.35) as follows:

$$[Zeo{\equiv}Al{-}OH]^-Na^+ + {}^-O{-}Si {\equiv} \rightleftarrows [Zeo{\equiv}Al{-}O{-}Si {\equiv}]^-Na^+ + OH^- \quad (2.34)$$

$$[Zeo{\equiv}Al{-}OH]^-Na^+ + HO{-}Si {\equiv} \rightleftarrows [Zeo{\equiv}Al{-}O{-}Si {\equiv}]^-Na^+ + H_2O \quad (2.35)$$

The incorporation of aluminum as a nucleophilic substitution reaction between the protonated silanol groups and the aluminate species solvated in the solution is described by Eq. 2.36 as follows:

$$Zeo{\equiv}Si{-}O^-Na^+ + Al(OH)_4^- \rightleftarrows [ZeoSi{-}O{-}Al(OH)_3]^-Na^+ + OH^- \quad (2.36)$$

Thus, aluminum as an electron acceptor via 3D orbitals has an important role in the incorporation of particles in crystal growth.

This model explains why the concentrations of silicon and aluminum in solution affect the growth rate (Bosnar et al. 2004). The performance of monomeric species in the synthesis of zeolites is evident in the satisfactory prediction of the solubility product of zeolite A, Π_s (Eq. 2.37) (Šefčík and McCormick 1997b) as follows:

$$\Pi_s = [Si(OH)_4][Al(OH)_4^-][Na^+] = 1, 2(\pm 0.3) \times 10^{-8} \ m^3 \quad (2.37)$$

(iii) Mineralizing agents

The function of the mineralizing agents is to contribute to the condensation equilibrium by converting the reactants into reactive and ionically motile species acting as catalysts (Cundy and Cox 2005). A mineralizing agent allows for the formation of a more stable solid phase from a less stable phase (Guth and Kessler 1999).

The most common sources of mineralizing agents are alkali metal hydroxides, hydrofluoric acid or fluorine salts.

The basic mechanism of action of the mineralizing agent consists of attacking the reagents with anions X^- (usually OH^- or F^-), forming derivative species AX^- (such as $Al(OH)_4^-$). After the pertinent reactions for the zeolite structure formation, the mineralizing anion agent is released and restarts the cyclic mechanism. Sometimes, these species are retained in the product (Cundy and Cox 2005).

(iv) Inorganic and organic cations

Cation sources are generally the same as mineralizing agents or alkali metal salts (Guth and Kessler 1999). Alkali metals are the most common zeolite charge compensators. According to Lee (1999), the structure of three-dimensional silicates tends to accommodate larger monoatomic cations because of their large interstices.

The function of cation structure drivers in formation is evident by the organization of the silicate and aluminosilicate species around these sources. The reactions in which a cation participates are exemplified by Eq. 2.38 (Do et al. 2014) as follows:

$$TO^- + Na^+ + HO-T \equiv\rightleftharpoons\equiv T-O^-Na^+ \cdots (OH)T \equiv\rightleftharpoons\equiv T-O-T \equiv +Na^+ + OH^- \quad (2.38)$$

The higher the ionic radius of the cation, the more silicon atoms are present in the single and double rings, and the presence of two different cations may change the degree of oligomerization of the species in solution (McCormick et al. 1987). In his study of zeolites as inclusion compounds, Bobonich (1994) concluded that the hydrated inorganic cation cluster/T-oxide tetrahedra molar ratio is an important factor to determine pore diameter.

Most of the zeolites synthesized in absence of organic compounds are obtained with Na^+ ions, but there are zeolites that are formed in the presence of other metallic ions (Yu 2007). The results of Gabelica et al. (1983) indicated that the size and variation in the morphology of crystals increased with the cationic radius.

Changes in the nature of alkaline inorganic cations do not necessarily change the nature of the silicate species in solution but moderately alter the silicate distribution (McCormick et al. 1987; Yu 2007) and interfere in the formation enthalpy of the zeolite networks (Navrotsky et al. 2009). Nonetheless, it is possible that zeolites with higher enthalpy of hydration have a direction of crystallization by endothermic effects of condensation of polymeric aluminosilicate species when there is variation in metallic cations with different hydrophilicity/hydrophobicity (Bobonich 1994).

Organic cations (space filler species, structural directing agents and templates) (Davis and Lobo 1992) interfere with the speciation of silicates (Šefčík and McCormick 1997a), even if their interactions are restricted to intermolecular forces (Cundy and Cox 2005). For example, the use of $N(CH_3)_4^+$ cations (TMA) in the synthesis of zeolite Y prevents zeolite P from being formed, leading to the subsequent decomposition to zeolite ZSM-4 (Cundy and Cox 2005). Organic cations promote the formation of zeolitic structures with lower charge density, increasing their framework stability (Lu et al. 2018).

The use of organic compounds helps to control the attributes of the zeolites, but for economy and the search for eco-friendly syntheses, routes that do not use organic compounds are encouraged (Maldonado et al. 2013; Mintova et al. 2016).

The presence of cations in the micropores reduces the energy required for the formation of the internal surfaces (Van Santen et al. 1986). The action of the cations on the formation of the zeolitic crystals is evident from the model of Lindner and Lechert (1994), where the role of Na^+ is essential, especially with the assumption of Do et al. (2014) that one of the driving forces in the process of crystallization is electrostatic.

Finally, when the structure is formed and presents an excess of electronic loads, certain cations remain in the structure, distributed according to the characteristics of the zeolitic phase (Yang 2003).

(v) Water

The importance of water is evident in hydrothermal synthesis and in other methods such as dry gel conversion, where water vapor promotes zeolite crystallization. In addition to acting as a solvent, water contributes to the changes that the reagents undergo in the process and participates directly in some of them, affecting the kinetics of the process and having large role in establishing the autogenous pressure (Yu 2007).

The amount of water must be controlled, as the concentrations of the species involved in the synthesis and consequently the supersaturation, the kinetics, the crystal size and even the phases to be formed depend on the amount of water (Yu 2007).

H_2O molecules react with the metal elements as Lewis bases via molecular orbital $3a_1$, forming σ bonds and generating OH_2, OH^- and O^{2-} ligands that are involved in the condensation reactions (Livage 1994).

Water also acts directly on the formation of zeolitic structures, because water molecules occupy channels and cavities in the structure. Through hydrogen bonds and hydrated hydroxyl ions that act as centers of clusterization, water is organized as compact clusters that can act by modeling the zeolitic structures (Bushuev et al. 2009).

The water content seems to influence the pore diameter, which affects the thermodynamic stability due to the inverse relationship of the water content with the opening of the structures (Giannetto et al. 2000; Byrappa and Yoshimura 2013).

Since ionic species act directly in the crystallization, the degree of purity of the water is controllable since the concentrations of the ionic species in the water can deeply influence the result of the synthesis, promoting the formation of undesired phases and interfering in the crystallization.

2.4.8 Time

Time is a common aspect in studies aimed at optimizing a synthesis. Zeolites are kinetically stable and tend to form thermodynamically stable structures over time (Zhang et al. 2016).

Although energetic aspects are related to crystal formation, no relationship has been found between the free energy and the formation of a specific zeolite phase, suggesting that kinetic factors take precedence in this domain (Ng et al. 2011).

Therefore, the kinetics are the most important aspects to be controlled. The parameters discussed above are related to the synthesis time; thus, variations in the parame-

Fig. 2.16 Crystallization
curves of zeolite A at
temperatures: (+) 50 °C, (◇)
60 °C, (○) 70 °C, (△) 80 °C
and (□) 90 °C. Reproduced
with permission from Park
et al. (2015), Copyright
(2015) Royal Society of
Chemistry

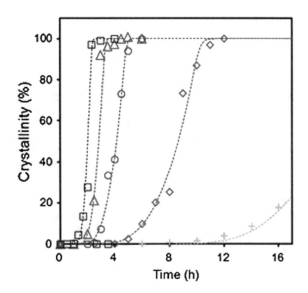

ters alters the optimum synthesis time (e.g., the higher the temperature, the less time is required for the appearance of a zeolite phase) (Fig. 2.16).

The crystal growth time around the initial core is determinant for the synthesis of a specific zeolite. This time can range from hours (zeolite LTA) to several days (zeolite MCM-22) (Mintova 2016). Synthesis formulations requiring less time are desirable.

From the industrial point of view, reduction in crystallization time is essential. According to Cundy and Cox (2005), the acceleration of synthesis should involve the shortening of the induction period, the transposition of a nonideal system to ideal conditions or a change in the balance between nucleation and crystallization.

It is common in synthesis studies to perform kinetic studies compatible with the purposes of the synthesis. These studies provide a certain understanding of the evolution of the reaction system towards a zeolitic phase and are important for observing the Ostwald rule of successive transformations, since for a system of a given composition with fixed rates of energy supply, different times may result in different phases in the final product by multicrystallization or the absence thereof.

Although crystallization cannot be adequately evaluated only by composition, time and temperature analysis (Wilson 2001), the nuances of synthesizing zeolites are useful in deeper understanding of the actions of these factors on the reaction system.

References

Aiello R, Barrer RM, Kerr LS (1974) Stages of zeolite growth from alkaline media. In: Flaningen EM, Sand LB (eds) Molecular Sieve Zeolites-I. Advances in chemistry series, vol 101. American Chemical Society, Washington, D.C., pp 44–50

Anderson MW, Gebbie-Rayet JT, Hill AR et al (2017) Predicting crystal growth via a unified kinetic three-dimensional partition model. Nature 544:456–459. https://doi.org/10.1038/nature21684

Angell CL, Flank WH (1977) Mechanism of zeolite A synthesis. In: Katzer JR (ed) Molecular sieves II. ACS symposium series, vol 40. American Chemical Society, Washington, D.C., pp 194–206

Barrer RM, Baynham JW, Bultitude FW, Meier WM (1959) 36. Hydrothermal chemistry of the silicates. Part VIII. Low-temperature crystal growth of aluminosilicates, and of some gallium and germanium analogues. J Chem Soc (Resumed) 195. https://doi.org/10.1039/jr9590000195

Bergna HE, Roberts WO (2006) Colloidal silica: fundamentals and applications. Taylor & Francis Group, LLC, Boca Raton, pp 9–36

Bobonich FM (1994) Crystallization of zeolites as formation of inclusion compounds. Theoret Exp Chem 30:106–116. https://doi.org/10.1007/bf00538189

Bosnar S, Subotić B (1999) Mechanism and kinetics of the growth of zeolite microcrystals. Microporous Mesoporous Mater 28:483–493. https://doi.org/10.1016/s1387-1811(98)00338-2

Bosnar S, Subotić B (2002) Kinect analysis of crystal growth of zeolite A. Croat Chem Acta 75(3):633–681. https://hrcak.srce.hr/127548. Accessed: Oct 2018

Bosnar S, Antonić T, Bronić J, Subotić B (2004) Mechanism and kinetics of the growth of zeolite microcrystals. Part 2: Influence of sodium ions concentration in the liquid phase on the growth kinetics of zeolite A microcrystals. Microporous Mesoporous Mater 76:157–165. https://doi.org/10.1016/j.micromeso.2004.07.021

Breck DW (1964) Crystalline molecular sieves. J Chem Educ 41:678. https://doi.org/10.1021/ed041p678

Brunner G (1992) A proposal for a mechanism of nucleation in zeolite synthesis. Zeolites 12:428–430. https://doi.org/10.1016/0144-2449(92)90042-n

Bushuev YG, Sastre G, Julián-Ortiz JVD (2009) The structural directing role of water and hydroxyl groups in the synthesis of beta zeolite polymorphs. J Phys Chem C 114:345–356. https://doi.org/10.1021/jp907694g

Byrappa K, Keerthinaj N, Byrappa SM (2015) Hydrothermal growth of crystals-design and processing. In: Rudolph P (ed) Handbook of crystal growth—bulk crystal growth: growth mechanisms and dynamics, vol 2A, 2nd edn. Elsevier, Amsterdam, pp 535–575

Byrappa K, Yoshimura M (2013) Handbook of hydrothermal technology, 2nd edn. Elsevier, Oxford

Caputo D, Gennaro B, Liguori B et al (2000) A preliminary investigation on kinetics of zeolite A crystallisation using optical diagnostics. Mater Chem Phys 66:120–125. https://doi.org/10.1016/s0254-0584(00)00310-2

Ciric J (1968) Kinetics of zeolite A crystallization. J Colloid Interface Sci 28:315–324. https://doi.org/10.1016/0021-9797(68)90135-5

Conato MT, Oleksiak MD, Mcgrail BP et al (2015) Framework stabilization of Si-rich LTA zeolite prepared in organic-free media. Chem Commun 51:269–272. https://doi.org/10.1039/c4cc07396g

Cook JD, Thompson RW (1988) Modeling the effect of gel aging. Zeolites 8:322–326. https://doi.org/10.1016/s0144-2449(88)80130-1

Corma A, Davis ME (2004) Issues in the synthesis of crystalline molecular sieves: towards the crystallization of low framework-density structures. ChemPhysChem 5:304–313. https://doi.org/10.1002/cphc.200300997

Cubillas P, Anderson MW (2010) Synthesis mechanism: crystal growth and nucleation. In: Čejka J, Corma AC, Zones SI (eds) Zeolites and catalysis: synthesis, reactions and applications, vol 1. Weinhiem: Willey-VCH Verlag GmbH & Co. KGaA, Weinhiem, pp 1–56

Cubillas P, Anderson MW, Attfield MP (2013) Materials discovery and crystal growth of zeolite A type zeolitic-imidazolate frameworks revealed by atomic force microscopy. Chem Eur J 19:8236–8243. https://doi.org/10.1002/chem.201300778

Culfaz A, Sand LB (1973) Mechanism of nucleation and crystallization of zeolites from Gels. In: Meier WM, Uytterhoeven JB (eds) Molecular sieves. Advances in chemistry series, vol 121. American Chemical Society, Washington, D.C., pp 140–151

Cundy CS, Cox PA (2003) The hydrothermal synthesis of zeolites: history and development from the earliest days to the present time. Chem Rev 103:663–702. https://doi.org/10.1021/cr020060i

Cundy CS, Cox PA (2005) The hydrothermal synthesis of zeolites: precursors, intermediates and reaction mechanism. Microporous Mesoporous Mater 82:1–78. https://doi.org/10.1016/j.micromeso.2005.02.016

Davies CW, Nancollas GH (1955) The precipitation of silver chloride from aqueous solutions. Part 3.—temperature coefficients of growth and solution. Trans Faraday Soc 51:818–823. https://doi.org/10.1039/tf9555100818

Davis ME, Lobo RF (1992) Zeolite and molecular sieve synthesis. Chem Mater 4:756–768. https://doi.org/10.1021/cm00022a005

Derouane EG, Determmerie S, Gabelica Z, Blom N (1981) Synthesis and characterization of ZSM-5 type zeolites I. physico-chemical properties of precursors and intermediates. Appl Catal 1:201–224. https://doi.org/10.1016/0166-9834(81)80007-3

Demazeau G (2010) Review. Solvothermal processes: definition, key factors governing the involved chemical reactions and new trends. Zeitschrift für Naturforschung B 65:999–1006. https://doi.org/10.1515/znb-2010-0805

Do MH, Wang T, Cheng D-G et al (2014) Zeolite growth by synergy between solution-mediated and solid-phase transformations. J Mater Chem A 2:14360. https://doi.org/10.1039/c4ta01737d

Dokter WH, Garderen HFV, Beelen TPM et al (1995) Homogeneous versus heterogeneous zeolite nucleation. Angew Chem Int Ed Engl 34:73–75. https://doi.org/10.1002/anie.199500731

Dupont J, Consorti CS, Spencer J (2000) Room temperature molten salts: neoteric "green" solvents for chemical reactions and processes. J Braz Chem Soc. https://doi.org/10.1590/s0103-50532000000400002

Fegan SG, Lowe BM (1986) Crystallisation of silicalite-1 precursors in the amine–$(C_3H_7)_4NBr$–SiO_2–H_2O system. J Chem Soc Faraday Trans 1: Phys Chem Condens Phases 82:801. https://doi.org/10.1039/f19868200801

Feijen EJP, Martens JA, Jacobs PA (1994) Zeolites and their mechanism of synthesis. In: Weitkamp J, Karge HG, Pfeifer H, Hölderich W (ed) Zeolites and related microporous materials. Studies in surface and science catalysis, vol 84A. Elsevier, Amsterdam, pp 3–21

Francis RJ, O'Hare D (1998) The kinetics and mechanisms of the crystallisation of microporous materials. J Chem Soc Dalton Trans 3133–3148. https://doi.org/10.1039/a802330a

Freund E (1976) Mechanism of the crystallization of zeolite X. J Cryst Growth 34:11–23. https://doi.org/10.1016/0022-0248(76)90257-8

Flörke OW et al (2005) Silica. In: Ullmann's Encyclopedia of industrial chemistry, vol 40, 7th edn. Wiley-VCH, Weinheim

Gabelica Z, Blom N, Derouane EG (1983) Synthesis and characterization of zsm-5 type zeolites. Appl Catal 5:227–248. https://doi.org/10.1016/0166-9834(83)80135-3

Gasteiger HA, Frederick WJ, Streisel RC (1992) Solubility of aluminosilicates in alkaline solutions and a thermodynamic equilibrium model. Ind Eng Chem Res 31:1183–1190. https://doi.org/10.1021/ie00004a031

Ghobarkar H, Schäf O, Massiani Y, Knauth P (2003) The reconstruction of Natural Zeolites. Dordrecht: Kluwer Academic Publishers. Cap. 1. p 1–6

Giannetto GP, Montes AR, Rodriguéz GF (2000) Zeolitas: características, propiedades y aplicaciones industriales. Caracas: Editorial Innovacín Tecnológica, Facultad de Ingeniería, UCV. p 351

Gies H, Marler B, Werthmann U (1998) Synthesis of porosils: crystalline nanoporous silicas with cage- and channel-like void structures. In: Karge H, Weitkamp J (eds) Synthesis. Molecular sieves science and technology series, vol 1. Springer, Heidelberg, pp 35–64

Gonthier S, Gora L, Güray I, Thompson RW (1993) Further comments on the role of autocatalytic nucleation in hydrothermal zeolite syntheses. Zeolites 13:414–418. https://doi.org/10.1016/0144-2449(93)90113-h

Grand J, Awala H, Mintova S (2016) Mechanism of zeolites crystal growth: new findings and open questions. CrystEngComm 18:650–664. https://doi.org/10.1039/c5ce02286j

Guth J-L, Kessler H (1999) Synthesis of aluminosilicate zeolites and related silica-based materials. In: Weitkamp J, Pupper L (eds) Catalysis and zeolites—fundamentals and applications. Springer, New York, pp 1–52

Hamilton KE, Coker EN, Sacco A et al (1993) The effects of the silica source on the crystallization of zeolite NaX. Zeolites 13:645–653. https://doi.org/10.1016/0144-2449(93)90137-r

Harrison PG, Hubberstey P (1974) Elements of group IV. In: Addison CC (ed) Inorganic Chemistry of the main-group elements, vol 2. The Chemical Society, London, pp 225–429

Howe-Grant M (1998) Kirk-Othmer Encyclopedia of Chemical Technology. 4 ed, vol 16. John Wiley & Sons, Inc

Kashchiev D, Rosmalen GMV (2003) Review: nucleation in solutions revisited. Cryst Res Technol 38:555–574. https://doi.org/10.1002/crat.200310070

Katz HS, Mileski JV (1987) Handbook of fillers for plastic. Van Nostrand Reinhold, New York

Kerr GT (1966) Chemistry of crystalline aluminosilicates. I. Factors affecting the formation of zeolite A. J Phys Chem 70:1047–1050. https://doi.org/10.1021/j100876a015

Kessler H (2001) Synthesis of high-silica zeolites and phosphate-based materials in the presence of fluoride. In: Robson HE (ed) Verified syntheses of zeolitic materials, 2nd edn. Elsevier, Amsterdam, pp 25–26

Knight CTG (1988) A two-dimensional Silicon-29 nuclear magnetic resonance spectroscopic study of the structure of the silicate anions present in an aqueous potassium silicate solution. J Chem Soc Dalton Trans 6:1457–1460. https://doi.org/10.1039/dt9880001457

Knight CTG, Wang J, Kinrade SD (2006) Do zeolite precursor species really exist in aqueous synthesis media? Phys Chem Chem Phys 8(26):3099–3103. https://doi.org/10.1039/B606419A

Knight CTG, Balec RJ, Kinrade SD (2007) The Structure of silicate anions in aqueous alkaline solutions. Angew Chem Int Ed 46(43):8148–8152. https://doi.org/10.1002/anie.200702986

Kühl G (2001) Source materials for zeolite synthesis. In: Robson HE (ed) Verified syntheses of zeolitic materials, 2nd edn. Elsevier, Amsterdam, pp 19–20

Lechert H (2001) The pH-value and its importance for the crystallization of zeolites. In: Robson HE (ed) Verified syntheses of zeolitic materials, 2nd edn. Elsevier, Amsterdam, pp 33–38

Lee JD (2009) Concise Inorganic Chemistry. Blackwell Science, Oxford. Portuguese edition: Lee JD (1999) Química Inorgânica não tão concisa (trans: Toma HE, Araki K, Rocha RC). Edgard Blucher, São Paulo

Leite ER, Ribeiro C (2012) Classical crystallization model: nucleation and growth. In: Leite ER, Ribeiro C (eds) Crystallization and growth of colloidal nanocrystals. Springer, New York, pp 19–44

Levine IN (2009) Physical chemistry, 6th edn. McGraw-Hill Higher Education, New York, Cap 4, pp 109–139

Li Q, Mihailova B, Creaser D, Sterte J (2001) Aging effects on the nucleation and crystallization kinetics of colloidal TPA-silicalite-1. Microporous Mesoporous Mater 43:51–59. https://doi.org/10.1016/s1387-1811(00)00346-2

Lim IH, Schrader W, Schüth F (2013) The formation of zeolites from solution—analysis by mass spectrometry. Microporous Mesoporous Mater 166:20–36. https://doi.org/10.1016/j.micromeso.2012.04.059

Lindner T, Lechert H (1994) The influence of fluoride on the crystallization kinetics of zeolite NaY. Zeolites 14:582–587. https://doi.org/10.1016/0144-2449(94)90194-5

Livage J (1994) Sol-gel chemistry and molecular sieve synthesis. In: Jansen JC et al (eds) Advanced zeolite science and applications. Studies in surface science and catalysis, vol 85. Elsevier, Amsterdam, pp 1–42

Louis B, Kiwi-Minsker L (2004) Synthesis of ZSM-5 zeolite in fluoride media: an innovative approach to tailor both crystal size and acidity. Microporous Mesoporous Mater 74:171–178. https://doi.org/10.1016/j.micromeso.2004.06.016

Lowe BM (1983) An equilibrium model for the crystallization of high silica zeolites. Zeolites 3:300–305. https://doi.org/10.1016/0144-2449(83)90173-2

Lu P, Villaescusa LA, Camblor MA (2018) Driving the crystallization of zeolites. Chem Rec 18:713–723. https://doi.org/10.1002/tcr.201700092

Lupulescu AI, Rimer JD (2014) In situ imaging of silicalite-1 surface growth reveals the mechanism of crystallization. Science 344:729–732. https://doi.org/10.1126/science.1250984

Madras G, McCoy BJ (2003) Temperature effects for crystal growth: a distribution kinetics approach. Acta Mater 51:2031–2040. https://doi.org/10.1016/s1359-6454(02)00621-3

Maldonado M, Oleksiak MD, Chinta S, Rimer JD (2013) Controlling crystal polymorphism in organic-free synthesis of Na-zeolites. J Am Chem Soc 135:2641–2652. https://doi.org/10.1021/ja3105939

Mathieu R, Vieillard P (2010) A predictive model for the enthalpies of formation of zeolites. Microporous Mesoporous Mater 132:335–351. https://doi.org/10.1016/j.micromeso.2010.03.011

McCormick AV, Bell AT (1989) The solution chemistry of zeolite precursors. Catal Rev 31:97–127. https://doi.org/10.1080/01614948909351349

McCormick AV, Bell AT, Radke CJ (1987) The effect of alkali metal cations on the structure of dissolved silicate oligomers. MRS Proc. https://doi.org/10.1557/proc-111-107

McNicol BD, Pott GT, Loos KR (1972) Spectroscopic studies of zeolite synthesis. J Phys Chem 76:3388–3390. https://doi.org/10.1021/j100667a016

McNicol BD, Pott GT, Loos KR, Mulder N (1973) Spectroscopic studies of zeolite synthesis: evidence for a solid-state mechanism. In: Meier WM, Uytterhoeven JB (eds) Molecular sieves. Advances in chemistry series, vol 121. American Chemical Society, Washington, D.C., pp 152–161

Meng X, Xiao F-S (2013) Green routes for synthesis of zeolites. Chem Rev 114:1521–1543. https://doi.org/10.1021/cr4001513

Mintova S (ed) (2016) Verified syntheses of zeolitic materials. XRD Patterns: Barrier N. 3rd rev. edn. Published on behalf of the Synthesis Commission of the International Zeolite Association 2016

Mintova S, Grand J, Valtchev V (2016) Nanosized zeolites: Quo Vadis? C R Chim 19:183–191. https://doi.org/10.1016/j.crci.2015.11.005

Mullin JW (2001) Crystallization, 4th edn. Butterworth-Heinemann, Oxford, pp 135–180

Navrotsky A, Trofymluk O, Levchenko AA (2009) Thermochemistry of microporous and mesoporous materials. Chem Rev 109:3885–3902. https://doi.org/10.1021/cr800495t

Ng E-P, Chateigner D, Bein T et al (2011) Capturing ultrasmall EMT zeolite from template-free systems. Science 335:70–73. https://doi.org/10.1126/science.1214798

Oleksiak MD, Rimer JD (2014) Synthesis of zeolites in the absence of organic structure-directing agents: factors governing crystal selection and polymorphism. Rev Chem Eng 30:1–49. https://doi.org/10.1515/revce-2013-0020

Park SH, Yang J-K, Kim J-H et al (2015) Eco-friendly synthesis of zeolite A from synthesis cakes prepared by removing the liquid phase of aged synthesis mixtures. Green Chem 17:3571–3578. https://doi.org/10.1039/c5gc00854a

Parnham ER, Morris RE (2007) Ionothermal synthesis of zeolites, metal-organic frameworks, and inorganic-organic hybrids. Acc Chem Res 40:1005–1013. https://doi.org/10.1021/ar700025k

Pérez-Pariente J (1995) Aspectos termodinámicos y cinéticos de la síntesis de zeolitas. In: Cardoso D, Urquieta-Gonzalez EA, Jahn SL (org) 2° Curso Iberoamericano sobre Peneiras Moleculares. CYTED. Rede Temática sobre Peneiras Moleculares São Carlos, pp 19–36

Petrovic I, Navrotsky A, Davis ME, Zones SI (1993) Thermochemical study of the stability of frameworks in high silica zeolites. Chem Mater 5:1805–1813. https://doi.org/10.1021/cm00036a019

Piccione PM, Yang S, Navrotsky A, Davis ME (2002) Thermodynamics of pure-silica molecular sieve synthesis. J Phys Chem B 106:3629–3638. https://doi.org/10.1021/jp014427j

Pope C (1998) Nucleation and growth theory in zeolite synthesis. Microporous Mesoporous Mater 21:333–336. https://doi.org/10.1016/s1387-1811(98)00016-x

Qin Z, Lakiss L, Tosheva L et al (2013) Comparative study of nano-ZSM-5 catalysts synthesized in OH⁻ and F⁻ media. Adv Func Mater 24:257–264. https://doi.org/10.1002/adfm.201301541

Qinhua X, Aizhen Y (1991) Hydrothermal synthesis and crystallization of zeolites. Prog Cryst Growth Charact Mater 21:29–70. https://doi.org/10.1016/0960-8974(91)90007-y

Randolph AD, Larson MA (1988) Theory of particulate processes, 2nd edn. Academic Press Inc., San Diego

Ren N, Subotić B, Bronić J et al (2012) Unusual pathway of crystallization of zeolite ZSM-5 in a heterogeneous system: phenomenology and starting considerations. Chem Mater 24:1726–1737. https://doi.org/10.1021/cm203194v

Sankar G, Okubo T, Fan W, Meneau F (2007) New insights into the formation of microporous materials by in situ scattering techniques. Faraday Discuss 136:157. https://doi.org/10.1039/b700090c

Schweizer M, Sagis LMC (2014) Nonequilibrium thermodynamics of nucleation. J Chem Phys 141:224102. https://doi.org/10.1063/1.4902885

Schwochow FE, Heinze GW (1974) Process of zeolite formation in the system. In: Flaningen EM, Sand LB (eds) Molecular sieves zeolites-I. Advanced in chemistry series, vol 101. American Chemical Society, Washington D.C., pp 102–108

Šefčík J, McCormick AV (1997a) Thermochemistry of aqueous silicate solution precursors to ceramics. AIChE J 43:2773–2784. https://doi.org/10.1002/aic.690431324

Šefčík J, McCormick A (1997b) What is the solubility of zeolite A? Microporous Mater 10:173–179. https://doi.org/10.1016/s0927-6513(97)00007-2

Šefčík J, McCormick A (1999) Prediction of crystallization diagrams for synthesis of zeolites. Chem Eng Sci 54:3513–3519. https://doi.org/10.1016/s0009-2509(98)00522-3

Sipos P (2009) The structure of Al(III) in strongly alkaline aluminate solutions—a review. J Mol Liq 146:1–14. https://doi.org/10.1016/j.molliq.2009.01.015

Subotić, B (1988) Influence of autocatalytic nucleation on zeolite crystallization processes. In: Occelli ML, Robson HE. (eds) Zeolite synthesis. ACS symposium series, vol 398. American Chemical Society, Washington D.C., pp 110–123

Subotić B, Graovac A (1985) Kinect analysis of autocatalytic nucleation during crystallization of zeolites. In: Držaj B, Hočevar S, Pejovnik S (eds) Zeolites, structure, technology and application. Proceedings of an international symposium organized by the "Boris Kidrič" Institute of Chemistry on behalf of the IZA, Portorož-Portorose, September 1984. (Studies in Surface Science and Catalysis), vol 24. Elsevier, Amsterdam, pp 199–206

Subotić B, Bronić J (2003) Theoretical and practical aspects of zeolite crystal growth. In: Auerbach SM, Carrado KA, Dutta PK (eds) Handbook of zeolite science and technology. Marcel Dekker Inc., New York, pp 125–199

Szostak R (1989) Molecular sieves: principles of synthesis and identification. Springer, New York, p 1989

Thompson RW (2001) Nucleation, growth, and seeding in zeolite synthesis. Verif Synth Zeolitic Mater 21–23. https://doi.org/10.1016/b978-044450703-7/50100-9

Valtchev V, Mintova S (2009) Nano/microporous materials: hydrothermal synthesis of zeolites. In: Lukehart CM, Scott RA (eds) Nanomaterials: inorganic and bioinorganic perspectives. Wiley, New York, pp 1–20

Van Santen RA, Keijsper J, Ooms G, Kortbeek AGTG (1986) The role of interfacial energy in zeoliete synthesis. In: Murakami Y, Iljima A, Ward JW (eds) New developments in zeolite science and technology. Proceedings of the 7th international zeolite conference, Tokyo (Studies in surface science and catalysis), vol 28. Elsevier, Amsterdam, pp 169–175

Van Speybroeck V, Hemelsoet K, Joos L et al (2015) Advances in theory and their application within the field of zeolite chemistry. Chem Soc Rev 44:7044–7111. https://doi.org/10.1039/c5cs00029g

Vinaches P, Bernardo-Gusmão K, Pergher S (2017) An introduction to zeolite synthesis using imidazolium-based cations as organic structure-directing agents. Molecules 22(8):1307

Wilson ST (2001) Templating in molecular sieve synthesis. In: Robson HE (ed) Verified synthesis of zeolitic materials, 2nd edn. Elsevier, Amsterdam, pp 27–31

Wu Z, Yang S, Wu W (2016) Shape control of inorganic nanoparticles from solution. Nanoscale 8:1237–1259. https://doi.org/10.1039/c5nr07681a

Xu R, Pang W, Yu J, Huo Q et al (2007) Chemistry of zeolites and related porous materials: synthesis and structure. Wiley, Singapore

Yang RT (2003) Adsorbents: fundamentals and applications. Wiley, New Jersey, pp 157–190

Yu J (2007) Synthesis of zeolites. In: Čejka J, Van Bekkum H, Corma A, Schüth F (eds) Introduction to zeolite science and practice. In: Studies in surface science and catalysis, vol 168, 3rd edn. Elsevier, Amsterdam, pp 39–103

Zhang H, Zhang H, Wang P et al (2016) Organic template-free synthesis of zeolite mordenite nanocrystals through exotic seed-assisted conversion. RSC Adv 6:47623–47631. https://doi.org/10.1039/c6ra08211d

Zhdanov SP (1974) Some problems of zeolite crystallization. In: Flaningen EM, Sand LB (eds) Molecular sieve zeolites-I. Advances in chemistry series, vol 101. American Chemical Society, Washington, D.C., pp 20–43

Chapter 3
Zeolite Eco-friendly Synthesis

In early 1980s, the knowledge of zeolite synthesis was already considerable, especially in relation to zeolite synthesis based on the use of inorganic structure-directing cations. At the same time, many studies began to explore the use of new sources of silicon and aluminum such as coal ash and ash from rice husk. The interest in these alternative sources of silicon and aluminum stemmed from two reasons: the first reason is to reduce the cost of zeolite synthesis of and the second reason is to add value to waste or natural raw materials such as clays. Adding value to waste is of great interest because these materials are usually environmental liabilities and using these materials in synthesis can generate profits and contribute to environmental remediation. This chapter addresses eco-friendly zeolite synthesis with the focus of using natural raw materials and waste. An overview of synthesis using these raw materials is presented, focusing on the main parameters studied in zeolite synthesis from alternative sources. The main advances in the syntheses using this type of material are presented.

3.1 Coal Ash Raw Materials

Coal still provides approximately 40% of the electricity consumed in the world according to data presented in World Energy Resources 2016 (World Energy Council 2016). According to the BP Statistical Review of World Energy 2017 (British Petroleum 2017), the proven reserves of coal in the world are currently sufficient to meet 153 years of global production. The Asia-Pacific region has the most proven reserves (46.5% of the total), with China accounting for 21.4% of the global total. The USA continues to be the largest holder of reserves (22.1% of total) (British Petroleum 2017). The production of ash from burning coal can be predicted to remain high and even become an bigger environmental problem despite advances in the optimization of coal burning.

Coal combustion products include ashes such as fly ash, bottom ash, boiler slag, fluidized bed combustion ash and flue gas desulfurization produced mainly by coal combustion or flue gas cleaning (Heidrich et al. 2013). Fly ashes are produced

© Springer Nature Switzerland AG 2019

R. Chaves Lima et al., *Environmentally Friendly Zeolites*, Engineering Materials,
https://doi.org/10.1007/978-3-030-19970-8_3

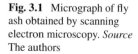

Fig. 3.1 Micrograph of fly ash obtained by scanning electron microscopy. *Source* The authors

in greater quantities, corresponding to approximately 85% of the ash produced. Figure 3.1 shows a micrograph of a sample of fly ash.

Fly ash has a high content of inorganic material at approximately 90–99%, which may be in amorphous or crystalline form (Vassilev and Vassileva 2005). Fly ashes are classified according to their application; Class C and Class F. Approximately 50–70% of Class C ash is Si, Al and Fe oxides, while more than 70% of Class F comprises these elements (Vassilev and Vassileva 2005). Class C ashes are produced from the burning of low grade coals (lignites or subbituminous coal), and class F ashes are produced from the burning of high grade coals (bituminous or anthracite coals) (Vassilev and Vassileva 2005; Ahmaruzzaman 2010).

Coal ashes are excellent sources of silicon and aluminum for zeolite synthesis, and knowledge regarding the use of these materials for this purpose is extensive. The first zeolites synthesized from coal ashes were zeolites with denser structures such as zeolites P, NaA (Stamboliev et al. 1985) and hydroxysodalite (Henmi 1987). In the 1990s, studies regarding zeolite synthesis using ashes as alternative sources began in earnest. Different zeolites can be synthesized from coal ash, for example, NaA (Wang et al. 2008; Bieseki et al. 2013a; Shih and Chang 1996), CaA, NaP1 (Kazemian et al. 2010; Cardoso et al. 2015), Li-ABW (Yao et al. 2009), Faujasite (Amrhein et al. 1996; Jin et al. 2015; Tanaka et al. 2002a), and CHA (Amrhein et al. 1996; Jin et al. 2015) zeolites.

In zeolite synthesis, different parameters and conditions can be modified to obtain the desired phase. We can divide the preparation process of zeolites from the ashes of coal into 3 different stages: the characterization of the starting materials, reaction and crystallization. In each of these steps, different parameters may be modified to obtain different zeolite phases or specifically a single phase. Figure 3.2 shows a diagram of the parameters involved and the specifics of each step.

Working with natural raw materials requires a thorough knowledge of each material that is replacing the conventional SiO_2 and Al_2O_3 sources. Thus, the first step of any synthesis based on alternative materials (including coal ashes) is character-

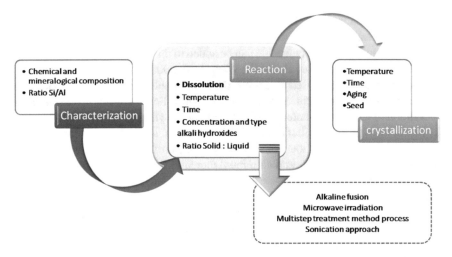

Fig. 3.2 Schematic view presents a diagram of the parameters involved and the specifics of each step. *Source* The authors

ization. Coal ashes may have different compositions including different types of metals, SiO_2/Al_2O_3 ratios and crystalline phases, which are approximately stable to the dissolution processes. Catalfamo et al. (1993) presented a very interesting study on the ash dissolution and crystallization of zeolitic phases. In this study, the presence of the mullite crystalline phase in the ash studied contributed to the stagnation in the crystallization process of the P and A zeolite phases (Catalfamo et al. 1993). These researchers presented dissolved Al and Si concentration curves and observed a significant increase in the Si concentration when the zeolite crystals appeared. The crystallization shutdown occurred because the aluminum available in the solution was consumed, and the rest of the aluminum present in the ashes was present in the form of mullite that was not dissolved (Catalfamo et al. 1993). Although ash is a material with different compositions, the preparation of pure zeolite phases is possible, mainly by adjusting the Si/Al ratio to values closer to those of conventional synthesis. This adjustment can be carried out by means of acid treatments or complementary addition of Si or Al, thus obtaining pure phases (Mondragon et al. 1990).

An important stage of reaction is process thinning synthesis using coal ash, and this process can be divided in stages. Previous studies show the importance of the dissolution of particles and assimilation of Si^{+4} and Al^{+3} for core formation (Catalfamo et al. 1993). Different strategies are used to improve the dissolution of the ashes. A strategy is set up in multistages or with different processes (Cardoso et al. 2015). In this type of synthesis, it is possible to synthesize more than one type of zeolite. The process consists of predissolving the ash in basic medium, and subsequently, the liquid and solids are separated by filtration. Al or Si is added to the supernatant to adjust the composition for the formation of a given zeolite, and the mixture begins the crystallization process. The solids collected from the first filtration may depend on the concentration of hydroxide used The solid may be zeolite NaP1 for example. If not, a certain amount of hydroxide can be added to this solid and undergo a process of crystallization (Tanaka et al. 2002b; Wang et al. 2008; Cardoso et al. 2015).

In relation to the proposed strategies, the ratio of solid: liquid is a key parameter for synthesis using coal ash. Higher ratios of solid: liquid are not favorable to the dissolution of the starting material (Wałek et al. 2008). There are correlations between the loss of mass during the first few hours of synthesis and the concentrations of Si and Al present; in the first few hours of synthesis, the mass of solid present decreases, and the concentrations of silicon and aluminum increase. With the formation of crystals, the mass increases, and the concentrations of both ions (Si^{+4} and Al^{+3}) decrease (Wałek et al. 2008).

Inorganic ions such as Na^+, K^+ and Ca^{+2} can act as inorganic structure directing agent. These cations are added in the form of hydroxides during synthesis. The type of hydroxide used may influence the type of structure formed, for example the use of KOH is required for the formation of K-chabazite zeolite. However, for reactions involving alkaline fusion, the use of KOH may not be efficient for the formation of zeolitic phases (Ríos et al. 2009). In a hydrothermal process using KOH, when the Si/Al ratio of the synthesis gel is approximately 1, changing the temperature of crystallization, time and liquid/ash ratio yields different types of zeolite structures such as CHA, EDI, LTF, SOD and PHI (Ríos et al. 2009).

Another parameter with great influence is the synthesis temperature; the higher the synthesis temperature, the denser the phases formed. This effect occurs because the water in the liquid phase solvates the inorganic cations, stabilizing the pores, and an increase in the synthesis temperature causes the water to transform to vapor phase, decreasing the efficiency of the water in stabilizing the pores (Byrappa and Yoshimura 2013). Regarding synthesis using coal ash, an increase in temperature can also lead to different types of structures (Lin and Hsi 1995). For Zhao et al. (1997), an increase in the synthesis temperature favored the synthesis of the P zeolite phase rather than the Y zeolite phase. A comparison between the crystallization curve of a conventional NaY zeolite and that of Y zeolite from ash showed that the process of crystallization using alternative sources is slower (Zhao et al. 1997). However, an increase in the aging temperature has a negative effect, since the temperature increase accelerates the process, and the more stable P zeolite phase is preferentially produced (Zhao et al. 1997). Hence, modifying parameters such as temperature and time does not produce the desired phase because many compounds present in coal ash are undesirable in synthesis, which has led certain studies to use acid treatments to remove these compounds (Mondragon et al. 1990).

Different studies use the alkali fusion process to dissolve coal ash to produce pure zeolitic phases (Ojha et al. 2004; Molina and Poole 2004; Kazemian et al. 2010; Hong et al. 2017; Hu et al. 2017). Generally, the temperature range studied or used in many works is approximately 400–650 °C using NaOH (Molina and Poole 2004; El-Naggar et al. 2008; Hong et al. 2017), KOH (Ríos et al. 2009) and LiOH. H_2O (Yao et al. 2009); however, with KOH, no zeolitic structure formed (Ríos et al. 2009). The alkaline fusion process favors the formation of Al and Si salts that are

Table 3.1 Data on syntheses performed using alkaline fusion

Zeolite	Alkaline fusion Temp/time	NaOH/Ash	Reaction Si/Al	Temp/time	Crystallization Temp/time	References
NaA	550 °C/1.5 h	1.5	1	nd	100 °C/12 h	Hong et al. (2017)
NaX	550 °C/1 h	1.0	1.8	Ab/24 h	90 °C/6 h	Molina and Poole (2004)
HS	550 °C/1 h	2.0	1.8	Ab/24 h	90 °C/6 h	Molina and Poole (2004)
NaX	550 °C/1 h	1.3	1.6	Ab/24 h	90 °C/6 h	Ojha et al. (2004)
NaP1	550 °C/1 h	1.2	1.1	70 °C/nd	120 °C/4 h	Kazemian et al. (2010)
NaX[a]	500 °C/1 h	1.2	1.25	Ab/12 h	100 °C/6 h	Shigemoto et al. (1993)
HS	500 °C/1 h	2.0	1.33	Ab/12 h	100 °C/6 h	Shigemoto et al. (1993)
SSZ-31	500 °C/1 h	1.5	1.2	Ab/16 h	80–100 °C/5–10 h	Sivalingam and Sen (2018)

Ab room temperature
[a]presence of traces of zeolite A

soluble, so most silicon and aluminum ions are available for synthesis. However, crystalline phases such as mullite may be more difficult to dissolve (El-Naggar et al. 2008). Higher melting temperatures are required when Na_2CO_3 is used in the process because Na_2CO_3 is less reactive than alkali metal hydroxides such as NaOH, but it is possible to obtain a material formed purely of aluminum and silica salts without the presence of quartz or mullite (Hu et al. 2017). The NaOH/ash ratio is also an important factor. Different relations NaOH/ash can produce different pure phases or mixtures. Table 3.1 presents the synthesis data of certain works cited in the literature that use of the stage of alkaline fusion.

Molina and Poole (2004) studied different NaOH/ash ratios under fixed conditions of synthesis: crystallization at 90 °C for 6 h with Si/Al = 1.8. These researchers observed that NaOH/ash reactions of 1.2 and 1.6 produced phase mixtures (zeolite X and hydroxysodalite). Ratios above 1.6 (NaOH/ash = 2) produced the pure hydroxysodalite phase, and for the ratio NaOH/ash = 1, pure X zeolite phase was produced (Molina and Poole 2004). Shigemoto et al. (1993) presented one of the first studies using alkaline fusion. In their studies, these researchers compared crystallinity values of NaA, NaX and HS zeolites in relation to the amounts of NaOH and ash used in the melting step. NaOH/ash ratios of 1–1.2 are favorable for NaX zeolite formation and ratios of 1.6–2 for HS zeolite. The maximum crystallinity for the NaX zeolite is obtained when the Si/Al ratio = 1.25 and the NaOH ratio/ash is 1.2 (Shigemoto et al. 1993). The use of a NaOH/ash ratio of 1.5 (Sivalingam and Sen 2018) for certain temperature and time conditions favored the formation of zeolite SSZ-31, a zeolite with a one-dimensional channel structure with elliptical pore apertures with openings of 12 Mb and sizes of 8.6 Å × 5.7 Å.

The syntheses of certain materials may take days, with crystallization processes using ash being many times slower than respective conventional syntheses of the desired zeolites. The use of microwave irradiation is an alternative method (Querol et al. 1997; Inada et al. 2005). One of the first studies using microwaves for synthesis with coal ash was carried out in (Querol et al. 1997), where the researchers observed a significant reduction in the synthesis time, from 24 h to 30 min, for the production of the zeolite phase NaP1. However, phases from the ash such as quartz, mullite and magnetite remained in low and medium concentrations. A further study of Inada et al. (2005) observed that microwave irradiation may inhibit the nucleation process, since accelerating the dissolution process prevents reprecipitation. This effect occurs because irradiation contributes to the dissolution of the nuclei, which are more unstable than the crystallites in the growth stage (Inada et al. 2005). The use of this type of radiation decreases the synthesis time when used in the dissolution stage for the Al and Si present in the ash. However, the use of this type of energy in the crystallization process is not favorable. Another alternative for alkaline fusion is the use of ultrasonic treatment (Belviso et al. 2011; Ojumu et al. 2016). Ojumu et al. (2016) showed that the fusion step can be replaced by 10 min of high intensity ultrasonic treatment.

As mentioned earlier, different zeolites can be produced using coal ash as feedstock. Many studies evaluate the feasibility of using these materials, mainly to applications directed to environmental remediation. Certain studies propose a cyclical use of materials, i.e., using materials produced from ash in the coal industry itself, such as in the remediation of highly polluting effluents. This idea is supported in multiple studies.

One of the more serious problems associated with mineral extraction is the generation of aqueous effluents when sulfide minerals are oxidized in the presence of water. These effluents, called acid mine drainage (AMD), are extremely harmful to the environment. A recent study Fallavena et al. (2018) showed that NaP1 and blackfill zeolite blends can be used to decrease the leaching of metals such as aluminum, iron, calcium, magnesium, zinc and manganese and the increase in pH caused by the presence of the zeolitic material.

3.2 Rusk Rice Ash Raw Materials

Ashes from rice husks are called tailings because these ashes are produced by burning the outer coating of the rice grains. The burning rice husks generates energy since most of the constituents of the husk are organic and only approximately 20% is silica; this composition varies from region to region and has trace amounts of elements such as Al, Fe, and Mg (Chandrasekhar et al. 2003). The ash produced by burning rice hulls has high silicon content, but if the synthesis is not controlled, lower silica contents and crystalline phases such as tridymite and cristobalite are obtained, making the silica produced less reactive (Chauhan and Kumar 2013). In a recent work, Costa and Paranhos conducted a study in 2018 regarding SiO_2 extraction from rice hull ash optimizing variables such as time, temperature and extractor concentration and proposing sustainable and clean production (Santana Costa and Paranhos 2018). With the combination of extraction with HCl and subsequent calcination at the temperature of 600 °C, excellent raw material for the synthesis of zeolite A can be obtained (Petkowicz et al. 2008).

The syntheses carried out using rice hull ash mainly prioritize the synthesis of materials with high SiO_2/Al_2O_3 ratios, for example, zeolites with ZSM-5 and mordenite structures; however, certain studies have also been related to the production of A, faujasite NaX and NaY and beta zeolites. The MOR type structure was synthesized using rice hull ash by Bajpai et al. (1981). The crystallization kinetics of the MOR type zeolite were very similar to those of conventional reagent synthesis. However, for synthesis with rice hull ash, a lower Na_2O/SiO_2 ratio is required compared with conventional synthesis, because by the crystallization process from rice husks ash is produced from a solution of reactive silica (Bajpai et al. 1981). Using ^{29}Si NMR techniques, a study conducted by Hamdan et al. (1997) showed that amorphous silica extracted by controlled-temperature physical combustion contained only tetrahedral units $*Si(OSi)_4$ which makes the material very reactive.

When working with alternative materials for zeolite synthesis, studies always seek to evaluate certain ranges of composition, temperature and time. However, a common premise is that all these parameters are close to values related to conventional syntheses, that is, to commercial raw materials. For synthesis of zeolite X using rice hull ash, Dalai et al. (1985) observed that the SiO_3/Al_2O_3 ratios for the zeolite varied from 3 to 5.5, in the range of conventional synthesis.

3.3 Clay Raw Materials

Clays are materials found in large quantities that are important raw materials for production sectors such as agriculture and industry. Clay can be defined as rock comprising different clay and nonclay minerals, organic matter and other impurities (Gomers 1988). When pulverized and mixed with a certain amount of water, clay becomes plastic, and after drying, clay becomes stiff. If clay is submitted to temperatures greater than 1000 °C, the materials acquires high hardness (Gomers 1988).

The different characteristics and properties of clays are attributed to different types of clay minerals that may be present in isolation or together with other minerals such as feldspar, mica and quartz (Gomers 1988; Santos 1989). These clay minerals can be described by structural units divided into two fundamental groups: tetrahedral and octahedral groups (Brigatti et al. 2006). Oxygen atoms or hydroxyl ions are arranged three-dimensionally around other atoms. In the tetrahedral arrangement, the central atoms are usually Si^{4+} and Al^{3+} and Mg^{2+}, Al^{3+}, Fe^{2+}, Ti^{4+} in octahedral groups (Santos 1989; Brigatti et al. 2006), Fig. 3.3.

Figure 3.3a shows the schematic representation of a tetrahedral sheet (a) and an octahedral sheet (b). In the tetrahedral sheet, the TO_4 units join through 3 vertices, sharing common oxygen atoms, basal oxygen, and forming a leaf with 6-membered rings (tetrahedrons) (Brigatti et al. 2006). Different from the tetrahedral units, the octahedral units bind to each other from the edges, presenting two possible topologies

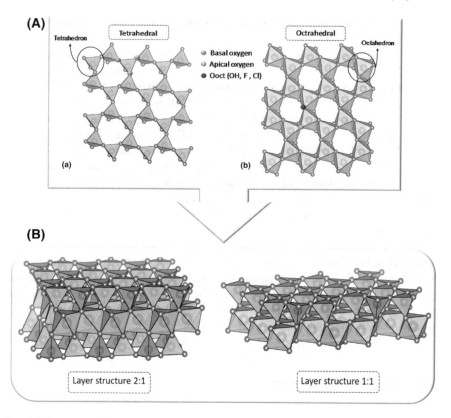

Fig. 3.3 Layer structures 1:1 and 2:1 formed from the union of octahedral and tetrahedral sheets. Layer structures 1:1 and 2:1 topology represented with Vesta (Momma and Izumi 2011)

in relation to the position of the OH (Oct) groups, *cis* and *trans* orientations (Brigatti et al. 2006). The junction between tetrahedral and octahedral sheets is a 1:1 sheet, and the junction of 2 tetrahedral sheets with an octahedral sheet is a 2:1 sheet. The connection between the tetrahedral sheet and the octahedral sheet occurs at the apical oxygen of the tetrahedral sheet (Brigatti et al. 2006). Representative schemes of the films are shown in Fig. 3.3b.

The way the leaves are organized is used to classify the clay minerals in structures: 1:1, 2:1 or mixed 2:1:1. The structures can also be subdivided into di- and tri-octahedral classifications according to the number of atoms in the structural unit that are in the octahedral sheet (Brigatti et al. 2006). In a 1:1 layer, the unit cell includes six octahedral sites and four tetrahedral sites. In the 2:1 structure, six octahedral sites and eight tetrahedral sites make up the unit cell. Structures with all six octahedral sites occupied are known as trioctahedral. If only four of the six octahedrons are occupied, the structure is called di-octahedral (Brigatti et al. 2006). The bonds that form the tetrahedral and octahedral leaves are strong bonds (partially ionic and covalent), but the connections between the leaves are weak and are responsible

Fig. 3.4 Micrograph of montmorillonite clay obtained by scanning electron microscopy. *Source* The authors

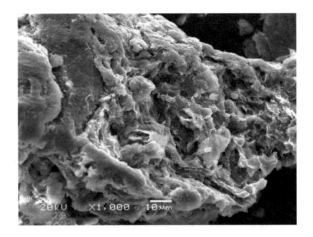

for easy cleavage parallel to the basal planes, which creates the morphological form of these materials (Santos 1989). A micrograph of a montmorillonite type clay is shown in Fig. 3.4.

In the 1980s, studies began using clays for zeolite synthesis. Kaolin is considered one of the most promising of these materials because it can produce different structures such as zeolites A, X and Y (Bosch et al. 1983; Costa et al. 1988). For synthesis using kaolin, the samples need to be calcined to produce a more reactive material called metakaolin. One of the first studies evaluating the calcination temperature of kaolin and the influence of kaolin on zeolite synthesis was conducted by Bosch et al. (1983), where these researchers observed that increasing calcination temperature (in the range of 750–1100 °C) resulted in decreasing crystallinity values for zeolite A, and from 950 °C, the presence of low zeolite Y crystallinity for both studied kaolins. An important finding appearing in this study and later in others was that unlike synthesis with conventional reagents, it is not possible to indiscriminately relate the composition of the clay used with the zeolite formed (Abdmeziem-Hamoudi and Siffert 1989; Tavasoli et al. 2014). For example, although the Si/Al ratio in many kaolins is similar to that of zeolite A, variations in the crystalline phases present often prevent all Si and Al being available for synthesis. These crystalline phases may be insoluble or partially soluble depending on the synthesis conditions and may be present as impurities in the kaolin or may be formed from the calcination step. Tavasoli et al. (2014) carried out a study of NaY zeolite synthesis using kaolin, evaluating important parameters such as calcination temperature, aging, time and crystallization temperature and chemical composition. Temperatures above 680 °C are not favorable because these temperatures produce mullite, a crystalline phase barely soluble in basic solutions that compromises the zeolite formation. The presence of other phases such as mullite is not always undesired; depending on the structural order of the kaolin at a given temperature, it is possible to obtain a larger amount of aluminum in a tetrahedral position, leading to an increase in the yield of zeolite produced, which can be evaluated by ^{27}Al MAS NMR (Maia et al. 2014, 2015).

Depending on the structure being synthesized, certain crystallization times can be very long; the use of microwaves can shorten this time. This effect occurs with syntheses using kaolin (Youssef et al. 2008).

Clays of type 2:1 such as montmorillonite have also been studied for the synthesis of zeolites. Using alkaline fusion as pretreatment and adjusting the Si/Al ratios, zeolite X can be produced (Musyoka et al. 2014); zeolite A (Ma et al. 2010) can be produced by also using Na_2CO_3 in the mixture that has undergone the melting process. Alkaline pretreatments similar to those occurring in the ash produce the Si^{+4} and Al^{+3} in solution and adjust the Si/Al ratios to values equal to or close to that of the finished zeolite, leading to purer and better phases (Abdmeziem-Hamoudi and Siffert 1989; Ma et al. 2010; Musyoka et al. 2014). The production of pure products with the use of smectite clays is more difficult than with kaolin. Many of the studies performed used this type of clay without previous treatment, and thus, the processes of synthesis become more delayed and mixtures of phases are obtained because Si and Al are not readily available for nucleation and subsequent crystal growth (Drag et al. 1985). However, for the formation of denser structures with pore systems with ring openings equal to or less than 8 MR, it is possible to produce pure phases such as SOD (sodalite) and GIS (zeolite NaP1) zeolites by varying the amount of NaOH in the synthesis system (Baccouche et al. 1998). Montmorillonite clay was also tested for ZSM-5 synthesis, but the products obtained showed low crystallinity (Mignoni et al. 2007). Illite clay is also a 2:1 clay, but illite belongs to the mica group. When a thermal treatment (550–800 °C) and subsequent acid treatment were carried out, it was possible to synthesize ZSM-5 without an organic driver using only aluminum sulfate as an external reagent (Han et al. 2019). The thermal treatment helps amorphize the structure, making the material more reactive, and the acid treatment removes large amounts of impurities such as Fe and Al. The high removal of Al allows the clay to be used as the main raw material for the synthesis of zeolites with high Si content, such as ZSM-5 (Han et al. 2019). Using an alkaline prefusion step, Mezni et al. (2011) observed effects of the Si/Al ratio, time of synthesis and temperature. These researchers observed that at 60 °C, pure zeolite X can be obtained in 24 h of synthesis, and at higher temperatures, the sodalite phase appears (Mezni et al. 2011).

3.4 Raw Powder Glass Materials

In the glass industry, not every residue can be reincorporated in the process because the quality of the residue can influence the fragility of the final product and the appearance of bubbles. The disposable waste is called raw powder glass, and the composition is fundamentally silicon-based. The morphology of this material is variable, as shown in Fig. 3.5.

The first reported reuse of these waste materials was published in 1997, regarding synthesized glass matrix composites (Boccaccini et al. 1997). For ten years, there were no new publications using these residues. In 2006, an article describing the

Fig. 3.5 Micrograph of raw powder glass obtained by scanning electron microscopy. *Source* The authors

DEMat-UFRN   2015/05/12   15:11 HL D6.6 ×3.0k   30 um

adsorption of heavy metals by this material was published (Catalfamo et al. 2006). However, in 2007 the most popular uses for this material appeared, incorporation in mortars and cements (Fragata et al. 2007). A new application was determined in 2014; the use of this material in traditional ceramics (Munhoz et al. 2014). Two last applications surged in 2018: the production of foam glass and incorporation in the pavement of cold regions (Rangel et al. 2018; Torabi Asl and Taherabadi 2018).

Glass residues were first used in the preparation of porous materials in 2012 and 2013, using waste bottle glass to synthesize ANA and GIS zeolites (Takei et al. 2012; Tsujiguchi et al. 2013). These zeolites do not have similar topologies, but they are usually found together in synthesis due to the Ostwald rule of successive transformations. The framework densities for the structures are 16.4 T/1000 $\mathring{A}^3$ and 19.2 T/1000 $\mathring{A}^3$ for GIS and ANA, respectively (Fig. 3.6) (Baerlocher et al. 2007). ANA zeolites have an ordered structure in the cubic crystallographic system. Meanwhile, GIS topology is classified in the tetragonal crystallographic system. However, both zeolites present 3-dimensional channel systems.

Industrial raw powder glass was incorporated for first time in zeolite synthesis in 2014 (Alves et al. 2014). This material is difficult to directly synthesize in basic media, so an alkaline fusion pretreatment was necessary in this study. Two pure phase topologies were obtained, SOD and CAN (Fig. 3.7). FAU and LTA zeolites were also found accompanying the SOD topology, as these structures have shared tiling, the 't-toc' (Fig. 3.8) (Baerlocher et al. 2007). These structures are cubic and have 3-dimensional channel systems. Contrary to the other topologies, the channel system of the CAN zeolite is 1-dimensional, and its crystallographic system is hexagonal.

In 2015, organic structure-directing agents were used in synthesis including recycling raw powder glass. The MFI topology (Fig. 3.9) was obtained, simulating the pure silica form (Vinaches et al. 2015). This zeolite is orthorhombic and has a 3-dimensional channel system. Surprisingly, another cation from the raw material, Fe^{3+}, was incorporated in the silicalite-1, which raised the possibility of synthesiz-

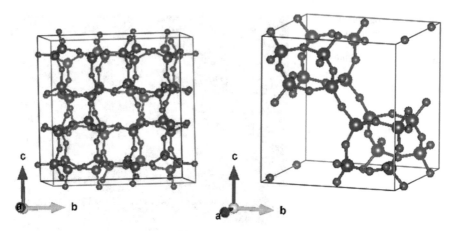

Fig. 3.6 ANA (left) and GIS (right) zeolitic topology represented with Vesta (Momma and Izumi 2011)

Fig. 3.7 CAN zeolitic topology represented with Vesta (Momma and Izumi 2011)

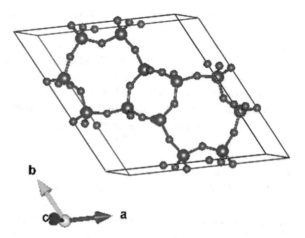

ing catalytic materials from this residue without needing other sources of cations to provide this characteristic.

Similarly, the synthesis of the MEL topology (Fig. 3.10) followed, also in the presence of a structure-directing organic (Vinaches et al. 2016). Both zeolites shared three composite building units, called *mor*, *mfi* and *mel*, but MFI also includes the *cas* composite building unit (Baerlocher et al. 2007). Both zeolites have 3-dimensional channel systems, although the MFI topology presents a helicoidal channel in one dimension. For the MEL synthesis, no unexpected heteroatoms were found in the framework.

A residue called waste glass cullet was recently used for zeolite synthesis (Majdinasab et al. 2019). In this study, microwave radiation was used, and a mixture of the ANA, GIS, LTA and SOD phases was obtained.

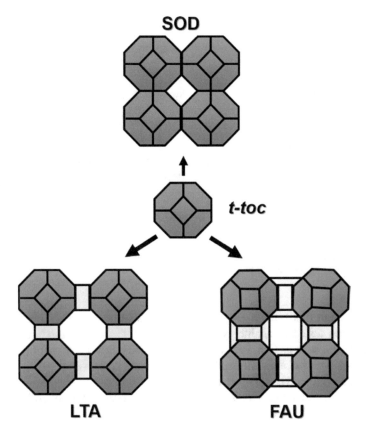

Fig. 3.8 Zeolitic topologies, SOD, LTA and FAU, obtained using raw powder glass, and their shared tiling, '*t-toc*', drawn in gray. The double four-membered cages are represented in blue, meanwhile the double six-membered rings are represented in pink. *Source* The authors

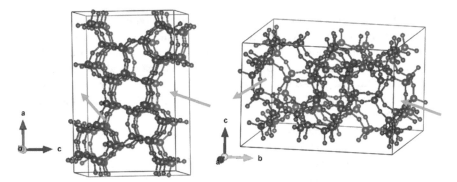

Fig. 3.9 MFI zeolitic topology represented with Vesta (Momma and Izumi 2011) in two different orientations showing the straight channels (left) and the sinusoidal channels (right)

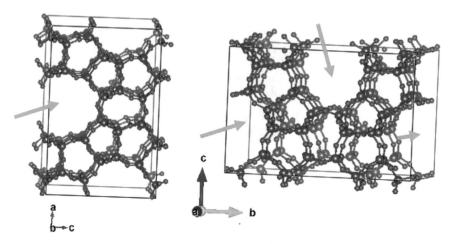

Fig. 3.10 MEL zeolitic topology represented with Vesta (Momma and Izumi 2011) in two different orientations showing the straight channels

Even though this section only discusses glass residues, newly obtained glass has also been tested in zeolite synthesis. Since 1995, zeolitic mixtures and pure zeolitic phases, such as GIS, PHI, SOD (Petrova and Kirov 1995), MFI (Dong et al. 1999), ANA (Dong et al. 1999), and FAU (Tatlier and Atalay-Oral 2017), have been synthesized using newly obtained glass.

This residue yielded very interesting and versatile results, and the zeolite synthesis recycling process requires further study to produce pure phases of other topologies.

3.5   Industrial Wastes as Raw Materials

In this consumer society, every day more and more products are produced, and industries do not stop until demand is reached. In parallel, the amounts of residues generated from production also continue growing. In this section, certain selected residues recovered in the form of zeolites are discussed, such as waste porcelain, red mud, electrolytic manganese residue, sugarcane bagasse and hazardous aluminum. The use of these residues as raw sources is very recent, so many possibilities are still available for research.

Waste porcelain. Residues from the ceramic industry have high proportions of silicon and aluminum. This material usually is recovered as part of new pottery (Seo et al. 2010), but this waste porcelain has also been used as part of concrete mixtures (Harini 2016) and for injection molding (Agote et al. 2001). The history of this material in zeolitic synthesis is recent. In 2007, the first attempt to convert this residue in zeolite was made, achieving phase mixtures with GIS and SOD topologies (Wajima and Ikegami 2007). The same researchers continued working with this material and successfully synthesized an FAU zeolite in 2009 (Wajima and Ikegami

2009). In 2014, a different research group published the synthesis of the LTA zeolite and proposed a catalytic application using the final product as support (Nezamzadeh-Ejhieh and Banan 2014). The three research articles all use alkaline fusion as a pretreatment. New studies developing novel methodologies for avoiding such an aggressive methodology are needed.

Red mud. Another industrial residue that has been recently studied is red mud. Red mud is a highly alkaline waste that contains high amounts of iron oxide and is mainly generated in the production of alumina, so this material is interesting as an aluminum source (López et al. 2011). The first application of this residue was its transformation to steel in 1971 (Guccione 1971). More specialized uses surged in 1976 when the potential of this material was studied for the hydrodemetallization of residual oils (Ueda et al. 1976). The catalytic properties of red mud and its use as a support are still currently exploited (Hamid et al. 2018). In 1982, Couillard published an article describing the adsorption capacity of in water treatment (Couillard 1982). Likewise, this property is still currently exploited (Deihimi et al. 2018).

The first attempt to use this residue as a part of zeolite synthesis was in 2015. Mixtures of LTA, FAU and SOD and mixtures of ANA and GIS were reported that year (Belviso et al. 2015; Zhao et al. 2016). Pure phases were achieved in 2018. LTA zeolite was obtained for CO oxidation and removal of heavy metals (Do Thi et al. 2018; Xie et al. 2018). The GIS topology was synthesized with this material to capitalize on the magnetic properties due to its composition (Belviso et al. 2018). These topologies were obtained in basic media in presence of Na^+ and showed the functionality of this residue as a raw source for zeolite synthesis. Likewise, a small number of pure phases were developed, highlighting the possibility for newer studies to recycle this waste.

Electrolytic manganese residue. This waste is generated in high quantities in the electrolytic manganese metal industry and is composed of silica, alumina and other oxides (Xin et al. 2011). The first attempt at recycling this waste was its incorporation in new building materials, published in 2010 (Wang et al. 2011). In the subsequent years, certain studies along this line of research were performed regarding the use of this material in cements (Hou et al. 2012; Yang et al. 2014) and bricks (Zhou et al. 2014).

In 2015, this waste was studied for zeolite synthesis, but only mixtures of CHA (Fig. 3.11), GIS and LTA topologies were obtained (Li et al. 2015a, b). Despite this result, the products were already used for dyes and heavy metals adsorption. The CHA topology is a trigonal crystallographic system with a 3-dimensional channel system (Baerlocher et al. 2007). This zeolite has a larger cavity size and is able to accommodate larger cations than GIS and LTA zeolites. This zeolite also has certain structural similarities to the CAN and GIS topologies.

The search for pure phase materials is a priority in the use of the electrolytic manganese residue.

Sugarcane bagasse. This residue is a major by-product of the sugar cane industry and is composed of cellulose, hemicellulose and lignin (Parameswaran 2009) but also contains silicon and aluminum. The first reuse reported in the literature was as an additive for cow rations in 1972 (Randel et al. 1972). Due to the composition of this

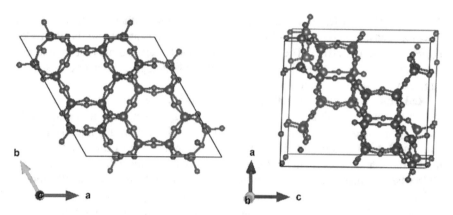

Fig. 3.11 CHA zeolitic topology represented with Vesta (Momma and Izumi 2011) in two different orientations

reside, researchers soon began to study its transformation in proteins and glucids (Srinivasan and Miller 1981). In 1983, the search for new alternative energy sources drove the use of this residue as fuel Elkoury (1983). As a result of the combustion process, in 1994 Girgis et al. reported the synthesis of sugarcane bagasse activated carbon (Girgis et al. 1994). A recent interest in greener fuels motivated researchers to continue studies on this waste, obtaining valuable products such as ethanol or biogas (Nosratpour et al. 2018).

Sugarcane bagasse was used in zeolite synthesis for first time in 2013 (Moisés et al. 2013). LTA zeolite was obtained as a pure phase. However, alkaline fusion was needed, demonstrating the need for further study to synthesize this zeolite in a 1-step reaction.

Hazardous aluminum. Aluminum can be recovered directly from bauxite or recycling waste materials (Galindo et al. 2015). Both processes generate residues that are used by another industry to recover the aluminum present in them. The residues from this industry are classified as hazardous as these residues are in the form of fine powder.

There is only one reference regarding the recovery of this product for zeolite synthesis, published in 2016 (Sánchez-Hernández et al. 2016). Sanchez-Hernandez et al. synthesized three pure phase products: ANA, SOD and GIS topologies.

3.6 Other Raw Materials

In this section, other natural materials used in zeolite synthesis that do not meet previous classifications are discussed, in particular, perlite, diatomite and spodumene.

Perlite. This material, shown in Fig. 3.12, is a natural volcanic material with the capacity to expand 7-16 times its volume when heated above 800 °C; the resultant material is called expanded perlite (da Silva Filho et al. 2017). The lamellar

Fig. 3.12 Micrograph of
expanded perlite from
scanning electron
microscopy. *Source* The
authors

morphology of perlite is responsible for its high external surface. The remarkable
proportions of silicon and aluminum in this material make it suitable as a nontra-
ditional raw source for zeolite synthesis. The first reference to this material was in
1876, when its structure was described. Currently, 4685 registers appear in Scopus
(October 22nd 2018), with 3536 registers published since 2000.

The first time this material was used in zeolite synthesis was in 1972 (Kirov
et al. 1979). Kirov et al. mixed perlite with NaOH + KOH solutions and hydrothermally synthesized mixtures of zeolitic topologies, such as ANA, MOR, HEU and
PHI (Fig. 3.13). MOR and PHI zeolites are orthorhombic, while the HEU topology is monoclinic (Baerlocher et al. 2007). Despite this difference, MOR and HEU
both have 2-dimensional channel systems. PHI zeolite has a 3-dimensional channel
system.

Even though perlite is a material that is difficult to dissolve, being almost stable
in hydrochloric acid media, certain pure phase topologies have been obtained. PHI
and SOD were obtained in 2004 by Rujiwatra using a perlite: sodium hydroxide ratio
of 1:5 Rujiwatra (2004). MFI and OFF (Fig. 3.14) topologies were also synthesized
in pure phases (Wang et al. 2007; da Silva Filho et al. 2018), but the presence of
unreacted perlite necessitates further study. The MFI synthesis described by Wang
et al. was also performed in basic media, adding MFI seeds to promote the structure-
direction in the absence of an organic compound. For the OFF topology, the effect
of tetramethylammonium hydroxide was evaluated, obtaining the desired topology
after 72 h. The OFF zeolite is characterized by a 3-dimensional channel system and
has a hexagonal structure (Baerlocher et al. 2007).

MOR and ANA topologies have also been reported as pure phase (Stenger et al.
1993; Tangkawanit et al. 2005), but the references did much not explore the synthesis
because the focus was on application. In conclusion, this natural material still has
much to offer as a substitute for traditional raw sources.

Diatomite. Another natural resource currently used as a silicon source is
diatomite. This material is composed of microscopic fossilized skeletons with vari-

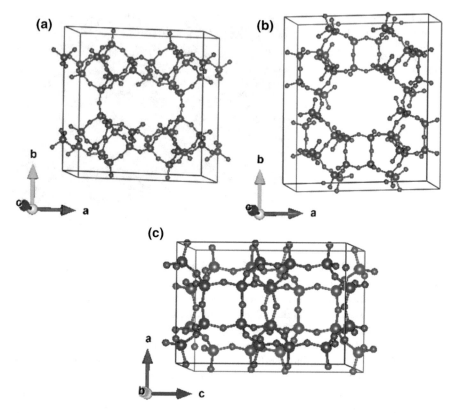

Fig. 3.13 HEU (**a**), MOR (**b**) and PHI (**c**) zeolitic topologies represented with Vesta (Momma and Izumi 2011)

Fig. 3.14 OFF zeolitic topology represented with Vesta (Momma and Izumi 2011)

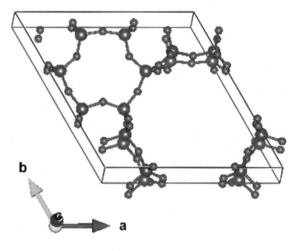

Fig. 3.15 Micrograph of diatomites obtained by scanning electron microscopy. *Source* The authors

DEMat-UFRN4932   2018/02/20 09:31 HL D7.1 x3.0k   30 um

able morphology (Fig. 3.15) (Danil de Namor et al. 2012). There are currently 3220 documents regarding diatomites found in Scopus (October 23rd 2018), but diatomites have been used since antiquity by the Greeks (Smol and Stoermer 2010). This material was rediscovered in 1836 by the studies of C. Fisher.

The first zeolitic product synthesized using diatomites as a raw source was the LTA topology in 1994 (Ghosh et al. 1994). At this time, diatomites were still treated as clays, but due to their precedence, diatomites are not part of this classification anymore. After 1994, this material was not used in zeolite synthesis until 2002. In 2003, Sanhueza et al. managed to obtain pure MOR and MFI topologies (Sanhueza et al. 2003). ANA, CAN, GIS and SOD, sodium-directed topologies were reported as pure phases obtained in 2005 (Chaisena and Rangsriwatananon 2005). Substitution of diatomites for traditional sources was also studied for the FAU, EDI and MER topologies (Fig. 3.16) in recent years (Novembre et al. 2014; Garcia et al. 2016). EDI and MER zeolites have tetrahedral crystallographic structures and 3-dimensional channel systems (Baerlocher et al. 2007). Their framework densities differ by 0.1, 16.3 tetrahedra/1000 Å^3 and 16.4 tetrahedra/1000 Å^3 for EDI and MER, respectively. The largest difference between these zeolites is the atomic disposition, with the EDI topology being a smaller pored zeolite than the MER topology.

Spodumene. This mineral, Fig. 3.17, is a lithium aluminosilicate found in a variety of colors, from colorless to purple to green, and the largest amount of this resource is located in Australia (Meshram et al. 2014). Spodumene is the main source of the lithium used in industry, containing approximately 3.7% of the metal. Through several pretreatments, it is possible to recover from 58 to 96% of the lithium, but a high quantity of residue remains. Thus, due to the composition of this waste, its reuse is interesting for zeolite synthesis.

Searching in Scopus for "spodumene" and "residue" together, only 16 documents appear (October 29th 2018). Most of these references are works about lithium extraction (Han et al. 2018). However, there is also a reference about zeolite synthesis by

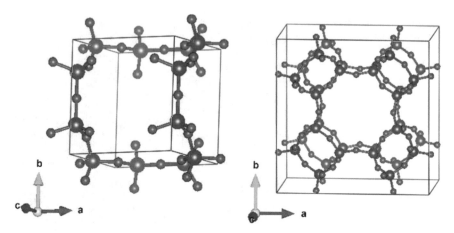

Fig. 3.16 EDI (left) and MER (right) zeolitic topologies represented with Vesta (Momma and Izumi 2011)

Fig. 3.17 Micrograph of spodumene obtained by scanning electron microscopy. *Source* The authors

Bieseki et al., reporting mixtures of GIS, LTA and SOD topologies (Bieseki et al. 2013b). The research group led by Sibele Pergher continued studying this residue and a M.Sc. dissertation and a Ph.D. thesis resulted from this study (Oliveira 2016; dos Santos 2018). These researchers reported the direct synthesis of pure phases of ABW, LTA and MER. Mixtures of CHA, FAU, GIS and MOR with unidentified impurities were also found. Pure phase EDI zeolite was also reported on (Oliveira and Pergher 2018). The ABW topology, in Fig. 3.18, was the only phase not reported previously, forming part of the orthorhombic crystallographic system and having a 1-dimensional channel system.

Fig. 3.18 ABW zeolitic topology represented with Vesta (Momma and Izumi 2011)

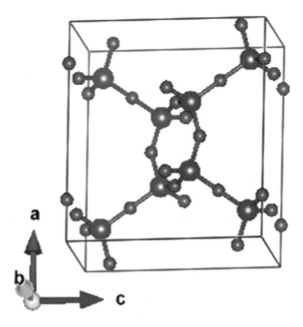

References

Abdmeziem-Hamoudi K, Siffert B (1989) Synthesis of molecular sieve zeolites from a smectite-type clay material. Appl Clay Sci 4:1–9. https://doi.org/10.1016/0169-1317(89)90010-0

Agote I, Odriozola A, Gutierrez M, Santamaría A, Quintanilla J, Coupelle P, Soares J (2001) Rheological study of waste porcelain feedstocks for injection moulding. J Eur Ceram Soc 21:2843–2853. https://doi.org/10.1016/s0955-2219(01)00210-2

Ahmaruzzaman M (2010) A review on the utilization of fly ash. Prog Energy Combust Sci 36:327–363. https://doi.org/10.1016/j.pecs.2009.11.003

Alves JABLR, Dantas ERS, Pergher SBC, Melo DMA, Melo MAF (2014) Synthesis of high value-added zeolitic materials using glass powder residue as a silica source. Mater Res 17:213–218. https://doi.org/10.1590/S1516-14392013005000191

Amrhein C, Haghnia GH, Kim TS, Mosher PA, Gagajena RC, Amanios T, De La Torre L (1996) Synthesis and properties of zeolites from coal fly ash. Environ Sci Technol 30:735–742. https://doi.org/10.1021/es940482c

Baccouche A, Srasra E, El Maaoui M (1998) Preparation of Na-P1 and sodalite octahydrate zeolites from interstratified illite-smectite. Appl Clay Sci 13:255–273. https://doi.org/10.1016/S0169-1317(98)00028-3

Baerlocher C, Mccusker LB, Olson DH (2007) Atlas of zeolite framework types. Elsevier B.V

Bajpai PK, Rao MS, Gokhale KVGK (1981) Synthesis of mordenite type zeolite using silica from rice husk ash. Ind Eng Chem Prod Res Dev 20:721–726. https://doi.org/10.1021/i300004a026

Belviso C, Cavalcante F, Lettino A, Fiore S (2011) Effects of ultrasonic treatment on zeolite synthesized from coal fly ash. Ultrason Sonochem 18:661–668. https://doi.org/10.1016/j.ultsonch.2010.08.011

Belviso C, Agostinelli E, Belviso S, Cavalcante F, Pascucci S, Peddis D, Varvaro G, Fiore S (2015) Synthesis of magnetic zeolite at low temperature using a waste material mixture: fly ash and red mud. Microporous Mesoporous Mater 202:208–216. https://doi.org/10.1016/j.micromeso.2014.09.059

Belviso C, Kharchenko A, Agostinelli E, Cavalcante F, Peddis D, Varvaro G, Yaacoub N, Mintova S (2018) Red mud as aluminium source for the synthesis of magnetic zeolite. Microporous Mesoporous Mater 270:24–29. https://doi.org/10.1016/j.micromeso.2018.04.038

Bieseki L, Penha FG, Pergher SBC (2013a) Zeolite A synthesis employing a brazilian coal ash as the silicon and aluminum source and its applications in adsorption and pigment formulation. Mater Res 16:38–43. https://doi.org/10.1590/s1516-14392012005000144

Bieseki L, Ribeiro DB, Sobrinho EV, Melo DMA, Pergher SBC (2013b) Synthesis of zeolites using silico-aluminous residue from the lithium extraction process. Ceramica 59:

Boccaccini AR, Bücker M, Bossert J, Marszalek K (1997) Glass matrix composites from coal flyash and waste glass. Waste Manag 17:39–45. https://doi.org/10.1016/S0956-053X(97)00035-4

Bosch P, Ortlz L, Schmer I (1983) Synthesis of faujasite type zeolites from calcined kaolins. Society 401–406

Brigatti MF, Galan E, Theng BKG (2006) Structures and Mineralogy of Clay Minerals. In: Bergaya F, Theng BKG, Lagaly G (eds) Handbook of clay science, pp 19–86

British Petroleum (2017) BP statistical review of world energy 2017. In: Br. Pet. https://www.bp.com/content/dam/bp/en/corporate/pdf/energy-economics/statistical-review-2017/bp-statistical-review-of-world-energy-2017-full-report.pdf. Accessed 11 Aug 2018

Byrappa K, Yoshimura M (2013) Hydrothermal synthesis and growth of zeolites. In: Handbook of hydrothermal technology, 2nd edn. Elsevier, pp 269–347

Cardoso AM, Horn MB, Ferret LS, Azevedo CMN, Pires M (2015) Integrated synthesis of zeolites 4A and Na-P1 using coal fly ash for application in the formulation of detergents and swine wastewater treatment. J Hazard Mater 287:69–77. https://doi.org/10.1016/j.jhazmat.2015.01.042

Catalfamo P, Corigliano F, Primerano P, Di Pasquale S (1993) Study of the pre-crystallization stage of hydrothermally treated amorphous aluminosilicates through the composition of the aqueous phase. J Chem Soc, Faraday Trans 89:171. https://doi.org/10.1039/ft9938900171

Catalfamo P, Primerano P, Arrigo I, Corigliano F (2006) Use of a glass residue in the removal of heavy metals from wastewater. Ann Chim 96:487–492. https://doi.org/10.1002/adic.200690049

Chaisena A, Rangsriwatananon K (2005) Synthesis of sodium zeolites from natural and modified diatomite. Mater Lett 59:1474–1479. https://doi.org/10.1016/j.matlet.2004.10.073

Chandrasekhar S, Satyanarayana KG, Pramada PN, Raghavan P, Gupta TN (2003) Processing, properties and applications of reactive silica from rice husk—an overview. J Mater Sci 38:3159–3168. https://doi.org/10.1023/A:1025157114800

Chauhan RP, Kumar A (2013) Radon resistant potential of concrete manufactured using ordinary portland cement blended with rice husk ash. Atmos Environ 81:413–420. https://doi.org/10.1016/j.atmosenv.2013.09.024

Costa E, de Lucas A, Uguina MA, Ruíz JC (1988) Synthesis of 4A zeolite from calcined kaolins for use in detergents. Ind Eng Chem Res 27:1291–1296

Couillard D (1982) Use of red mud, a residue of alumina production by the Bayer process, in water treatment. Sci Total Environ 25:181–191. https://doi.org/10.1016/0048-9697(82)90085-7

da Silva Filho SH, Vinaches P, Pergher SBC (2017) Caracterização estrutural da perlita expandida. Perspect Erechim 41:81–87

da Silva Filho SH, Vinaches P, Pergher SBC (2018) Zeolite synthesis in basic media using expanded perlite and its application in Rhodamine B adsorption. Mater Lett 227:258–260. https://doi.org/10.1016/j.matlet.2018.05.095

Dalai AK, Rao MS, Gokhale KVGK (1985) Synthesis of NaX zeolite using silica from rice husk ash. Ind Eng Chem Prod Res Dev 24:465–468. https://doi.org/10.1021/i300019a026

Danil de Namor AF, El Gamouz A, Frangie S, Martinez V, Valiente L, Webb OA (2012) Turning the volume down on heavy metals using tuned diatomite. A review of diatomite and modified diatomite for the extraction of heavy metals from water. J Hazard Mater 241–242:14–31. https://doi.org/10.1016/j.jhazmat.2012.09.030

Deihimi N, Irannajad M, Rezai B (2018) Equilibrium and kinetic studies of ferricyanide adsorption from aqueous solution by activated red mud. J Environ Manage 227:277–285. https://doi.org/10.1016/j.jenvman.2018.08.089

Do Thi MH, Tran QT, Nguyen T, Thi TVN, Huynh KPH (2018) Fabrication of CuO-doped catalytic material containing zeolite synthesized from red mud and rice husk ash for CO oxidation. Adv Nat Sci Nanosci Nanotechnol 9:025005

Dong W-Y, Sun Y-J, He H-Y, Long Y-C (1999) Synthesis and structural characterization of B-Al-ZSM-5 zeolite from boron–silicon porous glass in the vapor phase. Microporous Mesoporous Mater 32:93–100. https://doi.org/10.1016/S1387-1811(99)00094-3

dos Santos LL (2018) Extraction of lithium from beta-spodumene using routes with simultaneous acquisitium of zeolitic structures. Ph.D. theses, Federal University of Rio Grande do Norte

Drag EB, Abo-Lemon F, Rutkowski M, Miecznikowski A (1985) Synthesis of A, X and Y zeolites from clay minerals. Stud Surf Sci Catal 24:147–154. https://doi.org/10.1016/S0167-2991(08)65279-9

Elkoury JM (1983) Experimental electrical generating unit using sugarcane bagasse as fuel, pp 253–268

El-Naggar MR, El-Kamash AM, El-Dessouky MI, Ghonaim AK (2008) Two-step method for preparation of NaA-X zeolite blend from fly ash for removal of cesium ions. J Hazard Mater 154:963–72. https://doi.org/10.1016/j.jhazmat.2007.10.115

Fallavena VLV, Pires M, Ferrarini SF, Silveira APB (2018) Evaluation of zeolite/backfill blend for acid mine drainage remediation in coal mine. Energy Fuels 32:2019–2027. https://doi.org/10.1021/acs.energyfuels.7b03322

Fragata A, Paiva H, Velosa AL, Veiga MR, Ferreira VM (2007) Application of crushed glass residues in mortars, pp 923–927

Galindo R, Padilla I, Rodriguez O, Sanchez-Hernandez R, Lopez-Andres S, Lopez-Delgado A (2015) Characterization of solid wastes from aluminum tertiary sector: the current state of Spanish industry. J Miner Mater Charact Eng 3:55–64. https://doi.org/10.4236/jmmce.2015.32008

Garcia G, Cardenas E, Cabrera S, Hedlund J, Mouzon J (2016) Synthesis of zeolite Y from diatomite as silica source. Microporous Mesoporous Mater 219:29–37. https://doi.org/10.1016/j.micromeso.2015.07.015

Ghosh B, Agrawal DC, Bhatia S (1994) Synthesis of zeolite a from calcined diatomaceous clay: optimization studies. Ind Eng Chem Res 33:2107–2110. https://doi.org/10.1021/ie00033a013

Girgis BS, Khalil LB, Tawfik TAM (1994) Activated carbon from sugar cane bagasse by carbonization in the presence of inorganic acids. J Chem Technol Biotechnol 61:87–92. https://doi.org/10.1002/jctb.280610113

Gomers CF (1988) Argilas: O que são e para que servem. Fundação Galouste Gulbenkian, Lisboa

Guccione E (1971) "Red mud", a solid waste, can now be converted to high-quality steel. Eng Min J 172:136–138

Hamdan H, Muhid MNM, Endud S, Listiorini E, Ramli Z (1997) ^{29}Si MAS NMR, XRD and FESEM studies of rice husk silica for the synthesis of zeolites. J Non Cryst Solids 211:126–131. https://doi.org/10.1016/S0022-3093(96)00611-4

Hamid S, Bae S, Lee W (2018) Novel bimetallic catalyst supported by red mud for enhanced nitrate reduction. Chem Eng J 348:877–887. https://doi.org/10.1016/j.cej.2018.05.016

Han G, Gu D, Lin G, Cui Q, Wang H (2018) Recovery of lithium from a synthetic solution using spodumene leach residue. Hydrometallurgy 177:109–115. https://doi.org/10.1016/j.hydromet.2018.01.004

Han S, Liu Y, Yin C, Jiang N (2019) Fast synthesis of submicron ZSM-5 zeolite from leached illite clay using a seed-assisted method. Microporous Mesoporous Mater 275:223–228. https://doi.org/10.1016/j.micromeso.2018.08.028

Harini TA (2016) Experimental study on utilization of ceramic wastes in concrete. J Chem Pharm Sci 9:94–99

Heidrich C, Feuerborn H, Weir A (2013) Coal combustion products : a global perspective. World Coal Ash 17

Henmi T (1987) Synthesis of hydroxy-sodalite ("zeolite") from waste coal ash. Soil Sci Plant Nutr 33:517–521. https://doi.org/10.1080/00380768.1987.10557599

Hong JLX, Maneerung T, Koh SN, Kawi S, Wang C (2017) Conversion of coal fly ash into zeolite materials: synthesis and characterizations, process design, and its cost-benefit analysis. Ind Eng Chem Res 56:11565–11574. https://doi.org/10.1021/acs.iecr.7b02885

Hou P, Qian J, Wang Z, Deng C (2012) Production of quasi-sulfoaluminate cementitious materials with electrolytic manganese residue. Cem Concr Compos 34:248–254. https://doi.org/10.1016/j.cemconcomp.2011.10.003

Hu T, Gao W, Liu X, Zhang Y, Meng C (2017) Synthesis of zeolites Na-A and Na-X from tablet compressed and calcinated coal fly ash. R Soc Open Sci 4:170921. https://doi.org/10.1098/rsos.170921

Inada M, Tsujimoto H, Eguchi Y, Enomoto N, Hojo J (2005) Microwave-assisted zeolite synthesis from coal fly ash in hydrothermal process. Fuel 84:1482–1486. https://doi.org/10.1016/j.fuel.2005.02.002

Jin X, Ji N, Song C, Ma D, Yan G, Liu Q (2015) Synthesis of CHA zeolite using low cost coal fly ash. Procedia Eng 121:961–966. https://doi.org/10.1016/j.proeng.2015.09.063

Kazemian H, Naghdali Z, Ghaffari Kashani T, Farhadi F (2010) Conversion of high silicon fly ash to Na-P1 zeolite: alkaline fusion followed by hydrothermal crystallization. Adv Powder Technol 21:279–283. https://doi.org/10.1016/j.apt.2009.12.005

Kirov GN, Pechigargov V, Landzheva E (1979) Experimental crystallization of volcanic glasses in a thermal gradient field. Chem Geol 26:17–28. https://doi.org/10.1016/0009-2541(79)90027-5

Li C, Zhong H, Wang S, Xue J, Zhang Z (2015a) A novel conversion process for waste residue: Synthesis of zeolite from electrolytic manganese residue and its application to the removal of heavy metals. Colloids Surf A Physicochem Eng Asp 470:258–267. https://doi.org/10.1016/j.colsurfa.2015.02.003

Li C, Zhong H, Wang S, Xue J, Zhang Z (2015b) Removal of basic dye (methylene blue) from aqueous solution using zeolite synthesized from electrolytic manganese residue. J Ind Eng Chem 23:344–352. https://doi.org/10.1016/j.jiec.2014.08.038

Lin CF, Hsi HC (1995) Resource recoveiy of waste fly ash: synthesis of zeolite-like materials. Environ Sci Technol 29:1109–1117. https://doi.org/10.1021/es00004a033

López A, de Marco I, Caballero BM, Laresgoiti MF, Adrados A, Aranzabal A (2011) Catalytic pyrolysis of plastic wastes with two different types of catalysts: ZSM-5 zeolite and Red Mud. Appl Catal B Environ 104:211–219. https://doi.org/10.1016/j.apcatb.2011.03.030

Ma H, Ya Q, Fu Y, Ma C, Dong X (2010) Synthesis of zeolite of type a from bentonite by alkali fusion activation using Na_2CO_3. Ind Eng Chem Res 49:454–458. https://doi.org/10.1021/ie901205y

Maia AÁB, Angélica RS, de Freitas Neves R, Pöllmann H, Straub C, Saalwächter K (2014) Use of ^{29}Si and ^{27}Al MAS NMR to study thermal activation of kaolinites from Brazilian Amazon kaolin wastes. Appl Clay Sci 87:189–196. https://doi.org/10.1016/j.clay.2013.10.028

Maia AÁB, Neves RF, RôS Angélica, Pöllmann H (2015) Synthesis, optimisation and characterisation of the zeolite NaA using kaolin waste from the Amazon Region. Production of Zeolites KA, MgA and CaA. Appl Clay Sci 108:55–60. https://doi.org/10.1016/j.clay.2015.02.017

Majdinasab AR, Manna PK, Wroczynskyj Y, van Lierop J, Cicek N, Tranmer GK, Yuan Q (2019) Cost-effective zeolite synthesis from waste glass cullet using energy efficient microwave radiation. Mater Chem Phys 221:272–287. https://doi.org/10.1016/j.matchemphys.2018.09.057

Meshram P, Pandey BD, Mankhand TR (2014) Extraction of lithium from primary and secondary sources by pre-treatment, leaching and separation: a comprehensive review. Hydrometallurgy 150:192–208. https://doi.org/10.1016/j.hydromet.2014.10.012

Mezni M, Hamzaoui a., Hamdi N, Srasra E (2011) Synthesis of zeolites from the low-grade Tunisian natural illite by two different methods. Appl Clay Sci 52:209–218. https://doi.org/10.1016/j.clay.2011.02.017

Mignoni ML, Detoni C, Pergher SBC (2007) Estudo da síntese da zeólita ZSM-5 a partir de argilas naturais. Quim Nova 30:45–48. https://doi.org/10.1590/S0100-40422007000100010

Moisés MP, da Silva CTP, Meneguin JG, Girotto EM, Radovanovic E (2013) Synthesis of zeolite NaA from sugarcane bagasse ash. Mater Lett 108:243–246. https://doi.org/10.1016/j.matlet.2013.06.086

Molina A, Poole C (2004) A comparative study using two methods to produce zeolites from fly ash. Miner Eng 17:167–173. https://doi.org/10.1016/j.mineng.2003.10.025

Momma K, Izumi F (2011) VESTA 3 for three-dimensional visualization of crystal, volumetric and morphology data. J Appl Crystallogr 44:1272–1276

Mondragon F, Rincon F, Sierra L, Escobar J, Ramirez J, Fernandez J (1990) New perspectives for coal ash utilization: synthesis of zeolitic materials. Fuel 69:263–266. https://doi.org/10.1016/0016-2361(90)90187-U

Munhoz AH, Braunstein Faldini S, de Miranda LF, Masson TJ, Maeda CY, Zandonadi AR (2014) Recycling of automotive laminated waste glass in ceramic. Mater Sci Forum 798–799:588–593. https://doi.org/10.4028/www.scientific.net/MSF.798-799.588

Musyoka NM, Missengue R, Kusisakana M, Petrik LF (2014) Conversion of South African clays into high quality zeolites. Appl Clay Sci 97–98:182–186. https://doi.org/10.1016/j.clay.2014.05.026

Nezamzadeh-Ejhieh A, Banan Z (2014) Photodegradation of dimethyldisulfide by heterogeneous catalysis using nanoCdS and nanoCdO embedded on the zeolite A synthesized from waste porcelain. Desalin Water Treat 52:3328–3337. https://doi.org/10.1080/19443994.2013.797627

Nosratpour MJ, Karimi K, Sadeghi M (2018) Improvement of ethanol and biogas production from sugarcane bagasse using sodium alkaline pretreatments. J Environ Manage 226:329–339. https://doi.org/10.1016/j.jenvman.2018.08.058

Novembre D, Pace C, Gimeno D (2014) Syntheses and characterization of zeolites K-F and W type using a diatomite precursor. Mineral Mag 78:1209–1225. https://doi.org/10.1180/minmag.2014.078.5.08

Ojha K, Pradhan NC, Samanta AN (2004) Zeolite from fly ash: synthesis and characterization. Scanning 27:555–564

Ojumu TV, Du Plessis PW, Petrik LF (2016) Synthesis of zeolite A from coal fly ash using ultrasonic treatment—a replacement for fusion step. Ultrason Sonochem 31:342–349. https://doi.org/10.1016/j.ultsonch.2016.01.016

Oliveira MSM (2016) Síntese de zeólitas a partir de um resíduo sílico-aluminoso gerado na extração do lítio do espodumênio. MSc dissertation, Federal University of Rio Grande do Norte

Oliveira MSM, Pergher SBC (2018) Síntese de zeólita LPM-12 (tipo EDI) utilizando resíduo do processamento do espodumênio como fonte alternativa de silício e alumínio. Perspect, Erechim

Parameswaran B (2009) Sugarcane Bagasse. In: Singh nee' Nigam P, Pandey A (eds) Biotechnology for agro-industrial residues utilisation: utilisation of agro-residues. Springer Netherlands, Dordrecht, pp 239–252

Petkowicz D, Rigo R, Radtke C, Pergher S, Dossantos J (2008) Zeolite NaA from Brazilian chrysotile and rice husk. Microporous Mesoporous Mater 116:548–554. https://doi.org/10.1016/j.micromeso.2008.05.014

Petrova N, Kirov GN (1995) Zeolitization of glasses: a calorimetric study. Recent Adv Therm Anal Calorim 269–270:443–452. https://doi.org/10.1016/0040-6031(95)02353-4

Querol X, Alastuey A, López-Soler A, Plana F, Andrés JM, Juan R, Ferrer P, Ruiz CR (1997) A fast method for recycling fly ash: microwave-assisted zeolite synthesis. Environ Sci Technol 31:2527–2533. https://doi.org/10.1021/es960937t

Randel PF, Ramirez A, Carrero R, Valencia I (1972) Alkali-treated and raw sugarcane bagasse as roughages in complete rations for lactating cows. J Dairy Sci 55:1492–1495. https://doi.org/10.3168/jds.S0022-0302(72)85700-X

Rangel EM, de Melo CCN, C de O Carvalho, Osorio AG, Machado FM (2018) Espumas vítreas produzidas a partir de resíduos sólidos. Matéria 23:11967

Ríos RCA, Williams CD, Roberts CL (2009) A comparative study of two methods for the synthesis of fly ash-based sodium and potassium type zeolites. Fuel 88:1403–1416. https://doi.org/10.1016/j.fuel.2009.02.012

Rujiwatra A (2004) A selective preparation of phillipsite and sodalite from perlite. Mater Lett 58:2012–2015. https://doi.org/10.1016/j.matlet.2003.12.015

Sánchez-Hernández R, López-Delgado A, Padilla I, Galindo R, López-Andrés S (2016) One-step synthesis of NaP1, SOD and ANA from a hazardous aluminum solid waste. Microporous Mesoporous Mater 226:267–277. https://doi.org/10.1016/j.micromeso.2016.01.037

Sanhueza V, Kelm U, Cid R (2003) Synthesis of mordenite from diatomite: a case of zeolite synthesis from natural material. J Chem Technol Biotechnol 78:485–488. https://doi.org/10.1002/jctb.801

Santana Costa JA, Paranhos CM (2018) Systematic evaluation of amorphous silica production from rice husk ashes. J Clean Prod 192:688–697. https://doi.org/10.1016/j.jclepro.2018.05.028

Santos PDS (1989) Ciência e tecnologia de argilas, V.1, 2 edn. Edgar Bluncher Ltda, São Paulo

Seo DS, Han HG, Hwang KH, Lee JK (2010) Recycling of waste porcelain for ceramic ware. J Ceram Process Res 11:448–452

Shigemoto N, Hayashi H, Miyaura K (1993) Selective formation of Na-X zeolite from coal fly ash by fusion with sodium hydroxide prior to hydrothermal reaction. J Mater Sci 28:4781–4786. https://doi.org/10.1007/BF00414272

Shih WH, Chang HL (1996) Conversion of fly ash into zeolites for ion-exchange applications. Mater Lett 28:263–268. https://doi.org/10.1016/0167-577X(96)00064-X

Sivalingam S, Sen S (2018) Optimization of synthesis parameters and characterization of coal fly ash derived microporous zeolite X. Appl Surf Sci 455:903–910. https://doi.org/10.1016/j.apsusc.2018.05.222

Smol JP, Stoermer EF (2010) The diatoms: applications for the environmental and earth sciences. Cambridge University Press

Srinivasan VR, Miller TF (1981) Bioconversion of cellulose wastes to protein and sugar. Acta Biotechnol 1:365–370. https://doi.org/10.1002/abio.370010408

Stamboliev C, Scopova N, Bergk KH, Porsch M (1985) Synthesis of zeolite A and P from natural and waste materials. Stud Surf Sci Catal 24:155–160. https://doi.org/10.1016/S0167-2991(08)65280-5

Stenger HG, Hu K, Simpson DR (1993) Competitive adsorption of NO, SO_2 and H_2O onto mordenite synthesized from perlite. Gas Sep Purif 7:19–25. https://doi.org/10.1016/0950-4214(93)85015-N

Takei T, Ota H, Dong Q, Miura A, Yonesaki Y, Kumada N, Takahashi H (2012) Preparation of porous material from waste bottle glass by hydrothermal treatment. Ceram Int 38:2153–2157. https://doi.org/10.1016/j.ceramint.2011.10.057

Tanaka H, Furusawa S, Hino R (2002a) Synthesis, characterization, and formation process of Na–X zeolite from coal fly ash. J Mater 10:

Tanaka H, Sakai Y, Hino R (2002b) Formation of Na-A and -X zeolites from waste solutions in conversion of coal fly ash to zeolites. Mater Res Bull 37:1873–1884. https://doi.org/10.1016/s0025-5408(02)00861-0

Tangkawanit S, Rangsriwatananon K, Dyer A (2005) Ion exchange of Cu^{2+}, Ni^{2+}, Pb^{2+} and Zn^{2+} in analcime (ANA) synthesized from Thai perlite. Microporous Mesoporous Mater 79:171–175. https://doi.org/10.1016/j.micromeso.2004.10.040

Tatlier M, Atalay-Oral Ç (2017) Preparation of zeolite X coatings on soda-lime type glass plates. Brazilian J Chem Eng 34:203–210

Tavasoli M, Kazemian H, Sadjadi S, Tamizifar M (2014) Synthesis and characterization of zeolite nay using kaolin with different synthesis methods. Clays Clay Miner 62:508–518. https://doi.org/10.1346/CCMN.2014.0620605

Torabi Asl M, Taherabadi E (2018) Modification of silty clay strength in cold region's pavement using glass residue. Cold Reg Sci Technol 154:111–119. https://doi.org/10.1016/j.coldregions.2018.06.005

Tsujiguchi M, Kobashi T, Kanbara J, Utsumi Y, Kakimori N, Nakahira A (2013) Synthesis of zeolite from glass. J Soc Mater Sci Japan 62:357–361. https://doi.org/10.2472/jsms.62.357

Ueda S, Nakata Y, Yokoyama S-I, Ishii T (1976) Hydrodemetallization of residual oils with red mud catalyst. J Japan Pet Inst 19:234–238. https://doi.org/10.1627/jpi1958.19.234

Vassilev SV, Vassileva CG (2005) Methods for characterization of composition of fly ashes from coal-fired power stations: a critical overview. Energy Fuels 19:1084–1098. https://doi.org/10.1021/ef049694d

Vinaches P, Rebitski EP, Alves JABLR, Melo DMA, Pergher SBC (2015) Unconventional silica source employment in zeolite synthesis: raw powder glass in MFI synthesis case study. Mater Lett 159:233–236. https://doi.org/10.1016/j.matlet.2015.06.120

Vinaches P, Alves JABLR, Melo DMA, Pergher SBC (2016) Raw powder glass as a silica source in the synthesis of colloidal MEL zeolite. Mater Lett 178:217–220. https://doi.org/10.1016/j.matlet.2016.05.006

Wajima T, Ikegami Y (2007) Synthesis of zeolitic materials from waste porcelain at low temperature via a two-step alkali conversion. Ceram Int 33:1269–1274. https://doi.org/10.1016/j.ceramint.2006.05.020

Wajima T, Ikegami Y (2009) Synthesis of crystalline zeolite-13X from waste porcelain using alkali fusion. Ceram Int 35:2983–2986. https://doi.org/10.1016/j.ceramint.2009.03.014

Wałek TT, Saito F, Zhang Q (2008) The effect of low solid/liquid ratio on hydrothermal synthesis of zeolites from fly ash. Fuel 87:3194–3199. https://doi.org/10.1016/j.fuel.2008.06.006

Wang P, Shen B, Gao J (2007) Synthesis of ZSM-5 zeolite from expanded perlite and its catalytic performance in FCC gasoline aromatization. Catal Process Heavy Oil Upgrad 125:155–162. https://doi.org/10.1016/j.cattod.2007.03.010

Wang C-F, Li J-S, Wang L-J, Sun X-Y (2008) Influence of NaOH concentrations on synthesis of pure-form zeolite A from fly ash using two-stage method. J Hazard Mater 155:58–64. https://doi.org/10.1016/j.jhazmat.2007.11.028

Wang Y, Ye WH, Liu HB (2011) The research on preparation of a new building material with EMR and their properties. Adv Mater Res 163–167:4575–4579. https://doi.org/10.4028/www.scientific.net/AMR.163-167.4575

World Energy Council (2016) World energy resources. @WECouncil 2007:1–1028. http://www.worldenergy.org/wp-content/uploads/2013/09/Complete_WER_2013_Survey.pdf

Xie W-M, Zhou F-P, Bi X-L, Chen D-D, Li J, Sun S-Y, Liu J-Y, Chen X-Q (2018) Accelerated crystallization of magnetic 4A-zeolite synthesized from red mud for application in removal of mixed heavy metal ions. J Hazard Mater 358:441–449. https://doi.org/10.1016/j.jhazmat.2018.07.007

Xin B, Chen B, Duan N, Zhou C (2011) Extraction of manganese from electrolytic manganese residue by bioleaching. Bioresour Technol 102:1683–1687. https://doi.org/10.1016/j.biortech.2010.09.107

Yang C, Lv X, Tian X, Wang Y, Komarneni S (2014) An investigation on the use of electrolytic manganese residue as filler in sulfur concrete. Constr Build Mater 73:305–310. https://doi.org/10.1016/j.conbuildmat.2014.09.046

Yao ZT, Xia MS, Ye Y, Zhang L (2009) Synthesis of zeolite Li-ABW from fly ash by fusion method. J Hazard Mater 170:639–644. https://doi.org/10.1016/j.jhazmat.2009.05.018

Youssef H, Ibrahim D, Komarneni S (2008) Microwave-assisted versus conventional synthesis of zeolite A from metakaolinite. Microporous Mesoporous Mater 115:527–534. https://doi.org/10.1016/j.micromeso.2008.02.030

Zhao XS, Lu GQ, Zhu HY (1997) Effects of ageing and seeding on the formation of zeolite Y from coal fly ash. J Porous Mater 4:245–251. https://doi.org/10.1023/A:1009669104923

Zhao Y, Niu Y, Hu X, Xi B, Peng X, Liu W, Guan W, Wang L (2016) Removal of ammonium ions from aqueous solutions using zeolite synthesized from red mud. Desalin Water Treat 57:4720–4731. https://doi.org/10.1080/19443994.2014.1000382

Zhou C, Du B, Wang N, Chen Z (2014) Preparation and strength property of autoclaved bricks from electrolytic manganese residue. Spec Vol Sustain agenda Miner energy supply demand Netw an Integr Anal Ecol ethical, Econ Technol Dimens 84:707–714. https://doi.org/10.1016/j.jclepro.2014.01.052

Chapter 4
Recipes of Some Ecofriendly Syntheses

In this chapter, we described some ecofriendly synthesis studied and published by present and former members of LABPEMOL (Federal University of Rio Grande do Norte, Brazil), some of them in collaboration with other research groups. As described in the previous chapters, all the synthesis included in this section followed the hydrothermal synthesis (Fig. 4.1). The first step was referred to the gel formation, at room temperature (r. t.) or with heating. Then, the gel was transferred to an autoclave and introduced in a static oven at the required temperature for a variable number of days. And, finally, the solid product was recovered by filtration, washed and dried.

The synthesis following are classified depending on the natural material or residue added, ex. clay or raw powder glass. Every description is accompanied by a representation of the reference X-ray powder pattern from the IZA Database (Baerlocher and McCusker 2007) correspondent to the chosen topology. The characterization and complete description of the studies performed can be found in the published articles referenced in each subsection.

Fig. 4.1 Scheme of zeolite synthesis steps followed in the procedures described in this chapter

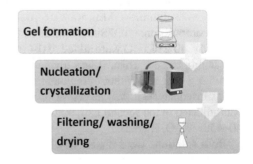

Gel formation

Nucleation/ crystallization

Filtering/ washing/ drying

© Springer Nature Switzerland AG 2019   93
R. Chaves Lima et al., *Environmentally Friendly Zeolites*, Engineering Materials,
https://doi.org/10.1007/978-3-030-19970-8_4

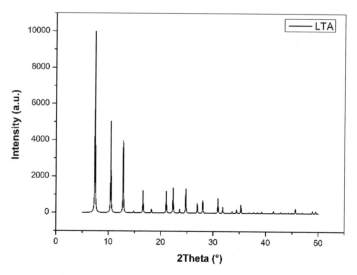

Fig. 4.2 Reference X-ray diffractogram of the LTA zeolite represented using the data from the IZA Database (Baerlocher and McCusker 2007)

4.1 Zeolites from Coal Ash Raw Material

Different types of zeolites are synthesized using coal ashes. In this subsection, we will describe A zeolite (LTA) synthesis (Bieseki et al. 2012), one of the most popular structures obtained using waste or alternative sources due to its industrial importance.

LTA zeolite Bieseki et al. (2012)

The synthesis chosen for this study used coal ash whose composition was 57.7% of SiO_2, 32.1% of Al_2O_3, 3.8% of K_2O and 3.7% of Fe_2O_3. The remainder elements present represented less than 1%. The synthesis of the zeolite was based in the reference IZA synthesis (Mintova 2016), describing a synthesis gel with the following molar composition: $0.037SiO_2$: $0.022Al_2O_3$: $0.071NaOH$: $2.40H_2O$.

Firstly, 40 g of H_2O were added to 0.4 g of NaOH (98%, Sigma) and the solution was homogenized. Then, it was divided into two halves. The first half was mixed with 3.9 g coal ash without previous treatment, 2.9 g of NaOH and 3.3 g of H_2O, and, subsequently, it was heated at 353 K for 2 h. Afterwards, the amount of Al was adjusted in the second half, adding 1.9 g of sodium aluminate (50–56% Al_2O_3, 40–45% Na_2O, Riedel).

Both solutions were mixed slowly and homogenized previously to the crystallization step at 373 K for 1 h, to avoid the presence of other impurities in the final product. After that time, the product was filtered, washed, dried, and characterized by X-ray diffraction as in Fig. 4.2 and Table 4.1.

Table 4.1 LTA topology: hkl datasheet corresponding to the 30 initial Bragg reflections from IZA Database (Baerlocher and McCusker 2007)

2 θ	d-spacing	Intensity	h	k	l
7.178	123.050	100.00	2	0	0
10.158	87.009	51.31	2	2	0
12.449	71.043	31.82	2	2	2
14.384	61.525	0.55	4	0	0
16.093	55.030	20.30	4	2	0
17.641	50.235	2.44	4	2	2
20.397	43.505	3.60	4	4	0
21.342	41.598	2.09	5	3	1
21.648	41.017	10.62	6	0	0
21.648	41.017	22.78	4	4	2
22.835	38.912	1.19	6	2	0
23.965	37.101	44.34	6	2	2
25.048	35.521	0.72	4	4	4
26.089	34.128	10.12	6	4	0
27.092	32.886	41.01	6	4	2
27.822	32.039	0.07	7	3	1
29.002	30.763	0.44	8	0	0
29.915	29.844	19.92	8	2	0
29.915	29.844	19.70	6	4	4
30.803	29.003	2.27	6	6	0
30.803	29.003	5.38	8	2	2
31.455	28.417	0.00	7	5	1
31.669	28.230	0.19	6	6	2
32.515	27.515	9.34	8	4	0
33.136	27.013	0.09	7	5	3
33.341	26.852	3.00	8	4	2
34.149	26.234	27.06	6	6	4
34.744	25.798	0.13	9	3	1
35.717	25.117	4.73	8	4	4
36.480	24.610	4.12	10	0	0

4.2   Zeolites from Rusk Rice Ash Raw Material

LTA structure can also be prepared from rusk rice ash (Petkowicz et al. 2008), dividing the synthesis procedure into two stages: a preparation of the ashes of paramount importance as mentioned earlier in the text, and the synthesis process itself. A step-by-step description of the reported synthesis procedure follows.

LTA zeolite Petkowicz et al. (2008)

Firstly, rice husk was submitted to acid leaching. This treatment was performed refluxing the residue in a 3% (v/v) hydrochloric acid (HCl, 37%, Merck) solution for 2 h at 100 °C. The ratio husk: solution was 50 g: 1 L. Afterwards, the leached husk was thoroughly washed with distilled water and then dried in an oven at 100 °C. After that, the sample was calcinated at 600 °C for 4 h.

Subsequently, the zeolite synthesis was performed, choosing a batch gel composition of $3.1Na_2O: Al_2O_3: 1.9SiO_2: 128H_2O$. The procedure was similar to the previously described with coal ash with small changes. An initial NaOH (98% Sigma) solution in water was divided in two equal volumes. In the first volume 1.6 g of sodium aluminate (50–56% Al_2O_3, 40–45% Na_2O, Riedel) was added and homogenized; and the second volume was mixed with 1.6 g of NaOH, 1.3 mL of H_2O and 0.9 g of husk rice ash. Then, the second volume solution was quickly poured into the first volume solution, stirring the mixture until homogenized. Afterwards, the gel was transferred to Teflon autoclaves (placed inside stainless steel autoclaves) for crystallization at 100 °C for 4 h. Finally, the sample was filtered, washed and dried, and its X-ray diffractogram was compared with the represented in Fig. 4.2 and Table 4.1.

4.3   Zeolites from Clay Raw Material

Clay materials have an important quantity of silicon and aluminium in their compositions. Two different classes, kaolin and montmorillonite, were employed in LABPE-MOL to obtain different zeolitic topologies. Here, we present LTA (Rigo et al. 2009), MFI (Mignoni et al. 2007) and MOR (Mignoni et al. 2008) zeolites synthesis.

LTA zeolite Rigo et al. (2009)

This zeolite was obtained using kaolin as an alternative silicon and aluminium source. The procedure followed was similar than the previous LTA synthesis using residues. In this case, all the silica was substituted with 8.2 g of kaolin (54.5% SiO_2, 34.3% Al_2O_3) and the Si/Al ratio was corrected adding 3.2 g of sodium aluminate (50–56% Al_2O_3, 40–45% Na_2O, Riedel). The time of synthesis needed to obtain a pure material was 4 h. As previously, the X-ray diffraction for comparison is found in Fig. 4.2 and Table 4.1.

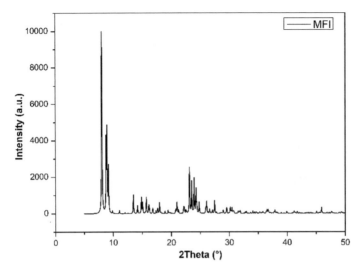

Fig. 4.3 Reference X-ray diffractogram of the MFI zeolite represented using the data from the IZA Database (Baerlocher and McCusker 2007)

MFI zeolite Mignoni et al. (2007)

MFI zeolite synthesis was studied with two different clays: kaolin (54.5% SiO_2, 34.3% Al_2O_3) and montmorillonite (60.9% SiO_2, 23.0% Al_2O_3). A solution containing 10 g of H_2O and 2.064 g of NaOH was prepared. To this solution is added 1.1 g of clay. Then 54.8 g H_2O and 0.24 g TPABr are added. As the co-director of structure 13.8 g ethyl alcohol is used. Finally, 8.3 g of SiO_2 is added and the gel is kept stirring for 30 min. The gel thus prepared was sustained in autoclaves at 150 °C for 3 days.

The X-ray diffractograms obtained were compared with the represented in Fig. 4.3 and Table 4.2.

MOR zeolite Mignoni et al. (2008)

This zeolite was reported with kaolin as alternative raw material. As in the previous procedures, the synthesis gel was initially defined. In the MOR synthesis, the molar ratio composition was $30SiO_2$: Al_2O_3: $6Na_2O$: $780H_2O$. As in the case of the LTA synthesis, the first step involved the preparation of a NaOH solution, in this case weighing 4.7 g of NaOH (98%, Sigma) and 10 g of H_2O. Afterwards, the kaolin (54.5% SiO_2, 34.3% Al_2O_3) was added, adjusting the Si/Al ratio with sodium aluminate (50–56% Al_2O_3, 40–45% Na_2O, Riedel). Once homogenized, 161.3 g of H_2O were added, and the mixture was stirred for 30 min. Then, 5 mg of commercial mordenite were added into the gel to act as seeds. Finally, the crystallization was performed at 170 °C for 1 day. The final product was filtered, washed and dried, and its X-ray diffractogram was compared with the presented in Fig. 4.4 and Table 4.3.

Table 4.2 MFI topology: hkl datasheet corresponding to the 30 initial Bragg reflections from IZA Database (Baerlocher and McCusker 2007)

2θ	d-spacing	Intensity	h	k	l
7.940	111.263	68.40	1	0	1
7.955	111.051	50.25	0	1	1
8.826	100.110	31.28	2	0	0
8.880	99.495	36.75	0	2	0
9.099	97.114	25.40	1	1	1
9.882	89.430	4.68	2	1	0
11.028	80.163	1.02	2	0	1
11.892	74.357	2.76	2	1	1
11.923	74.166	10.40	1	2	1
12.533	70.570	5.44	2	2	0
13.220	66.915	6.79	0	0	2
13.943	63.464	17.38	1	0	2
14.176	62.423	0.35	2	2	1
14.638	60.464	8.56	1	1	2
14.820	59.725	10.91	3	0	1
14.894	59.431	4.11	0	3	1
15.477	57.204	0.85	3	1	1
15.540	56.974	10.04	1	3	1
15.918	55.632	6.20	2	0	2
15.948	55.526	5.59	0	2	2
16.015	55.294	3.54	2	3	0
16.532	53.577	1.77	2	1	2
16.554	53.506	1.50	1	2	2
17.303	51.208	3.54	3	2	1
17.338	51.104	0.53	2	3	1
17.704	50.055	2.60	4	0	0
17.815	49.748	5.55	0	4	0
18.255	48.557	0.04	2	2	2
18.261	48.543	0.48	4	1	0
18.763	47.254	0.00	3	0	2

Table 4.3   MOR topology: hkl datasheet corresponding to the 30 initial Bragg reflections from IZA Database (Baerlocher and McCusker 2007)

2 θ	d-spacing	Intensity	h	k	l
6.503	135.811	100.00	1	1	0
8.607	102.650	14.31	0	2	0
9.760	90.550	59.44	2	0	0
13.027	67.906	0.35	2	2	0
13.437	65.842	44.32	1	1	1
13.822	64.015	33.97	1	3	0
14.580	60.705	16.74	0	2	1
15.286	57.915	12.11	3	1	0
17.263	51.325	0.67	0	4	0
17.574	50.423	2.87	2	2	1
18.176	48.767	5.35	1	3	1
19.321	45.903	3.49	3	1	1
19.591	45.275	1.82	4	0	0
19.593	45.270	26.75	3	3	0
19.868	44.651	0.70	2	4	0
20.931	42.407	1.51	0	4	1
21.433	41.425	6.46	4	2	0
22.181	40.044	55.16	1	5	0
22.904	38.796	1.29	3	3	1
23.141	38.404	20.48	2	4	1
23.617	37.640	8.01	0	0	2
24.507	36.293	5.42	4	2	1
24.521	36.273	0.38	1	1	2
24.943	35.669	2.71	5	1	0
25.169	35.353	0.40	1	5	1
25.179	35.339	5.17	0	2	2
25.608	34.757	91.76	2	0	2
26.020	34.217	5.93	0	6	0
26.226	33.953	0.00	4	4	0
26.227	33.951	53.10	3	5	0

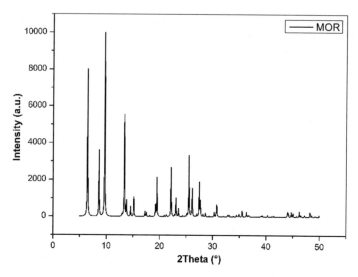

Fig. 4.4 Reference X-ray diffractogram of the MOR zeolite represented using the data from the IZA Database (Baerlocher and McCusker 2007)

4.4 Zeolites from Raw Powder Glass Materials

In this subsection, only the obtention of isolated zeolitic topologies are described. Firstly, it appears the syntheses of cancrinite (CAN zeolitic topology) and sodalite (SOD zeolitic topology) involving an alkaline fusion pre-treatment and a basic mineralizer. Both syntheses were reported in the article by Alves et al. (2014). MFI and MEL zeolites syntheses follow, obtained directly using hydrofluoric media as mineralizer, and reported in 2015 (Vinaches et al. 2015) and in 2016 (Vinaches et al. 2016) respectively.

CAN zeolite Alves et al. (2014)

The reagents used were raw powder glass (70.5% SiO_2, 1.2% Al_2O_3), aluminum oxide (pure, Sigma-Aldrich), sodium hydroxide (97%, Sigma-Aldrich) and distilled water. The quantities needed from each reagent were calculated adjusting the final synthesis gel molar ratio to $3SiO_2: Al_2O_3: 5.7 Na_2O: 228H_2O$.

Firstly, raw powder glass, alumina and sodium hydroxide were weighed in a porcelain crucible and heated at 350 °C for 2 h. Then, distilled water was added to the resulting product and mixed until a gel was formed. This gel was placed in a Teflon autoclave, and then, into a stainless-steel autoclave. To obtain a pure phase CAN zeolite, the aging can be performed either at 152 or 170 °C for 1 day. The final product was filtered, washed and dried. Its resulting X-ray powder pattern can be compared with the presented in Fig. 4.5 and Table 4.4.

Table 4.4   CAN topology: hkl datasheet corresponding to the 30 initial Bragg reflections from IZA Database (Baerlocher and McCusker 2007)

2 θ	d-spacing	Intensity	h	k	l
8.073	109.422	2.86	1	0	0
14.007	63.175	62.18	1	1	0
16.187	54.711	1.45	2	0	0
19.138	46.337	100.00	1	0	1
21.468	41.358	0.97	1	2	0
21.468	41.358	6.82	2	1	0
22.345	39.754	0.70	1	1	1
23.794	37.364	4.30	2	0	1
24.384	36.474	64.55	3	0	0
27.716	32.160	60.92	1	2	1
27.716	32.160	57.15	2	1	1
28.228	31.588	0.32	2	2	0
29.407	30.348	2.07	1	3	0
29.407	30.348	1.75	3	1	0
30.066	29.697	6.06	3	0	1
32.709	27.356	45.09	4	0	0
33.310	26.876	0.03	2	2	1
34.330	26.100	18.19	1	3	1
34.330	26.100	14.04	3	1	1
35.058	25.575	44.08	0	0	2
35.738	25.103	1.48	2	3	0
35.738	25.103	0.90	3	2	0
36.034	24.904	1.19	1	0	2
37.244	24.122	35.98	4	0	1
37.639	23.878	1.88	1	4	0
37.639	23.878	1.18	4	1	0
37.923	23.706	1.28	1	1	2
38.837	23.169	1.82	2	0	2
39.974	22.536	9.31	2	3	1
39.974	22.536	10.89	3	2	1

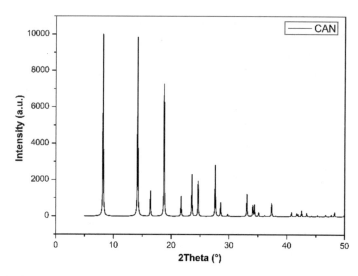

Fig. 4.5 Reference X-ray diffractogram of the CAN zeolite represented using the data from the IZA Database (Baerlocher and McCusker 2007)

SOD zeolite Alves et al. (2014)

In the referenced article, two synthesis variations were described that produced SOD zeolite. Most part of the synthesis was maintained, being identical to the previous for CAN zeolite: the reagents and synthesis gel molar ratio were unchanged.

The first variation decreased the alkaline fusion temperature and the aging temperature to 250 and 100 °C respectively. Meanwhile, in the second variation, the only change was a decrease in the aging temperature to 135 °C.

The reference X-ray powder pattern in both cases can be compared to the shown in Fig. 4.6 and Table 4.5.

MFI zeolite Vinaches et al. (2015)

In this research work the reagents employed were raw powder glass residues (76.3% SiO_2), tetrapropylammonium hydroxide (TPAOH, 1 M in water, Sigma-Aldrich), hydrofluoric acid (HF, 40% weight, Cinética) and deionized water. All the reagents, except HF, were weighed and magnetically stirred for 2 h. Then, HF was added, and the mixture was mixed for 30 min. The final molar ratio composition of the synthesis gel was SiO_2: 0.5TPAOH: 0.5 HF: $40H_2O$. In this case, the aging was performed at 180 °C for 10 days, and the resulting product presented an X-ray diffractogram coincident with the reference (Fig. 4.3 and Table 4.2).

MEL zeolite Vinaches et al. (2016)

The reagents were the same than in the previous synthesis but TPAOH, which was replaced by tetrabutylammonium hydroxide (TBAOH, 40% in water, Fluka). The synthetic procedure was also repeated with minor changes in the reagents' quantities

Table 4.5 SOD topology: hkl datasheet corresponding to the 30 initial Bragg reflections from IZA
Database (Baerlocher and McCusker 2007)

2 θ	d-spacing	Intensity	h	k	l
14.144	62.565	53.18	1	1	0
20.054	44.240	15.24	2	0	0
22.450	39.569	3.64	2	1	0
24.625	36.122	100.00	2	1	1
28.510	31.282	19.78	2	2	0
31.960	27.980	98.07	3	1	0
35.104	25.542	77.60	2	2	2
36.587	24.540	0.07	3	2	0
38.021	23.647	13.65	3	2	1
40.758	22.120	2.90	4	0	0
42.071	21.460	0.07	4	1	0
43.351	20.855	16.12	3	3	0
43.351	20.855	19.89	4	1	1
45.826	19.785	4.97	4	2	0
47.024	19.308	0.12	4	2	1
48.200	18.864	7.55	3	3	2
50.490	18.061	7.01	4	2	2
51.607	17.696	0.03	4	3	0
52.707	17.352	4.64	5	1	0
52.707	17.352	9.44	4	3	1
55.915	16.430	0.06	5	2	0
55.915	16.430	0.59	4	3	2
56.957	16.154	5.23	5	2	1
59.006	15.641	39.30	4	4	0
61.012	15.174	1.52	5	3	0
61.012	15.174	17.52	4	3	3
62.000	14.956	0.17	5	3	1
62.979	14.747	2.48	6	0	0
62.979	14.747	5.56	4	4	2
63.950	14.546	0.08	6	1	0

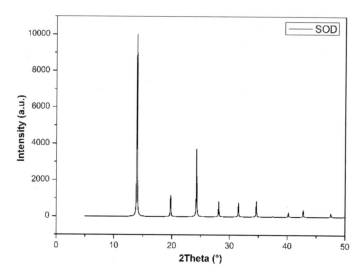

Fig. 4.6 Reference X-ray diffractogram of the SOD zeolite represented using the data from the IZA Database (Baerlocher and McCusker 2007)

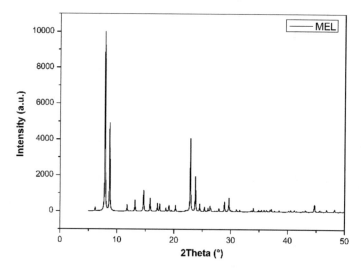

Fig. 4.7 Reference X-ray diffractogram of the MEL zeolite represented using the data from the IZA Database (Baerlocher and McCusker 2007)

and the aging temperature. For this experiment, the molar ratio composition of the final synthesis gel was SiO_2: 0.35TBAOH: 0.35 HF: $12H_2O$; and the aging temperature was 114 °C for 21 days. The X-ray diffractogram of the final product can be compared with the presented in Fig. 4.7 and Table 4.6.

Table 4.6 MEL topology: hkl datasheet corresponding to the 30 initial Bragg reflections from IZA Database (Baerlocher and McCusker 2007)

2 θ	d-spacing	Intensity	h	k	l
6.224	141.895	2.38	1	1	0
7.923	111.502	100.00	1	0	1
8.806	100.335	53.65	2	0	0
11.856	74.584	3.90	2	1	1
12.466	70.948	0.37	2	2	0
13.193	67.055	9.25	0	0	2
13.944	63.457	0.00	3	1	0
14.599	60.626	0.64	1	1	2
14.787	59.858	15.49	3	0	1
15.883	55.751	8.00	2	0	2
17.236	51.405	3.56	3	2	1
17.664	50.168	7.07	4	0	0
18.189	48.733	0.06	2	2	2
18.745	47.298	1.63	3	3	0
19.241	46.091	3.70	3	1	2
19.386	45.750	0.30	4	1	1
19.769	44.871	0.18	4	2	0
20.336	43.634	3.21	1	0	3
22.111	40.170	0.20	4	0	2
22.198	40.014	0.59	2	1	3
22.575	39.355	1.14	5	1	0
22.991	38.651	1.17	3	3	2
23.113	38.449	52.35	5	0	1
23.113	38.449	0.28	4	3	1
23.841	37.292	0.12	4	2	2
23.922	37.167	22.73	3	0	3
24.777	35.903	5.13	5	2	1
25.082	35.474	0.47	4	4	0
25.537	34.853	2.59	3	2	3
25.867	34.415	0.12	5	3	0

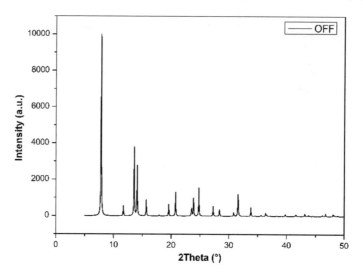

Fig. 4.8 Reference X-ray diffractogram of the OFF zeolite represented using the data from the IZA Database (Baerlocher and McCusker 2007)

4.5   Zeolites from Other Raw Materials

This subsection includes the synthesis of two zeolitic materials, OFF (da Silva Filho et al. 2018) and EDI (Oliveira et al. 2018) topologies, synthesized from perlite and spodumene, respectively.

OFF zeolite da Silva Filho et al. (2018)

The reagents used in this research work were 10.3 g of expanded perlite (Si/Al = 4.2), 29.3 g of tetramethylammonium hydroxide (TMAOH, 25%, Sigma-Aldrich), 1.6 g of sodium hydroxide (97%, Sigma-Aldrich) and 29.0 g of deionized water. Firstly, sodium hydroxide was dissolved in deionized water. Then, expanded perlite and TMAOH were added and the solution was stirred for 1 h previously to the aging step. Once the gel was formed and introduced in the autoclave, the aging was performed at 100 °C for 3 days. Finally, the product was filtered, washed, dried, and characterized by X-ray diffraction as in Fig. 4.8 and Table 4.7.

EDI zeolite Oliveira et al. (2018)

this research work the reagents were spodumene residue (Si/Al = 3.3), aluminium triisopropylate (98%, Merck), potassium hydroxide (KOH, 85%, Vetec) and deionized water. Firstly, an 8.2 M KOH solution in deionized water was prepared. Then, this solution was divided between V1 and V2. 1.0 g of aluminium source was added to V1 and stirred; meanwhile 8.8 g of the residue were mixed with V2. Once the different volumes were homogenized for 10 min, V2 was poured in V1 and the resulting solution was stirred for 20 min, until gel formation. Subsequently, the gel

Table 4.7   OFF topology: hkl datasheet corresponding to the 30 initial Bragg reflections from IZA Database (Baerlocher and McCusker 2007)

2 θ	d-spacing	Intensity	h	k	l
7.674	115.103	100.00	1	0	0
11.662	75.820	38.72	0	0	1
13.312	66.455	14.23	1	1	0
13.975	63.318	11.53	1	0	1
15.383	57.552	12.12	2	0	0
17.733	49.976	0.32	1	1	1
19.347	45.841	14.73	2	0	1
20.397	43.505	33.87	2	1	0
23.163	38.368	6.73	3	0	0
23.447	37.910	1.50	0	0	2
23.557	37.734	43.93	2	1	1
24.705	36.007	33.35	1	0	2
26.006	34.234	1.46	3	0	1
26.808	33.228	1.81	2	2	0
27.056	32.929	1.97	1	1	2
27.925	31.924	4.45	3	1	0
28.164	31.659	15.93	2	0	2
29.323	30.433	0.84	2	2	1
30.354	29.422	5.75	3	1	1
31.053	28.776	19.29	4	0	0
31.270	28.581	33.31	2	1	2
33.194	26.967	2.44	3	0	2
33.275	26.903	4.96	4	0	1
33.920	26.407	0.31	3	2	0
35.490	25.273	0.64	0	0	3
35.717	25.118	10.35	4	1	0
35.909	24.988	0.17	2	2	2
35.984	24.937	4.01	3	2	1
36.364	24.685	0.26	1	0	3
36.775	24.419	0.75	3	1	2

was transferred to a Teflon autoclave, and then, placed in a stainless-steel autoclave. The aging was performed at 95 °C for 68 h. Finally, the zeolite was filtered, washed and dried. The result obtained by X-ray characterization can be compared with the presented in Fig. 4.9 and Table 4.8.

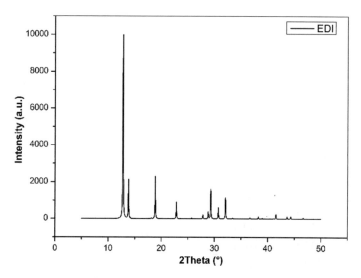

Fig. 4.9 Reference X-ray diffractogram of the EDI zeolite represented using the data from the IZA Database (Baerlocher and McCusker 2007)

Table 4.8 EDI topology: hkl datasheet corresponding to the 30 initial Bragg reflections from IZA Database (Baerlocher and McCusker 2007)

2 θ	d-spacing	Intensity	h	k	l
13.040	67.836	1.36	1	1	0
13.593	65.090	90.99	0	0	1
16.413	53.964	34.93	0	1	1
16.475	53.762	46.26	1	0	1
18.370	48.255	53.91	0	2	0
18.592	47.685	43.83	2	0	0
18.879	46.966	86.41	1	1	1
20.611	43.057	8.98	1	2	0
20.760	42.751	4.05	2	1	0
22.923	38.764	5.46	0	2	1
23.103	38.467	1.55	2	0	1
24.772	35.911	100.00	1	2	1
24.897	35.733	86.71	2	1	1

(continued)

Table 4.8 (continued)

2 θ	d-spacing	Intensity	h	k	l
26.253	33.918	20.70	2	2	0
27.382	32.545	10.88	0	0	2
28.929	30.839	14.99	0	1	2
28.965	30.801	13.27	1	0	2
29.274	30.483	29.47	1	3	0
29.560	30.194	30.31	3	1	0
29.676	30.079	20.74	2	2	1
30.438	29.343	35.24	1	1	2
30.982	28.840	1.22	0	3	1
31.288	28.565	1.52	3	0	1
32.405	27.605	84.11	1	3	1
32.666	27.391	90.87	3	1	1
33.175	26.982	0.41	0	2	2
33.303	26.881	0.58	2	0	2
33.576	26.669	5.11	2	3	0
33.735	26.547	5.88	3	2	0
34.517	25.963	41.08	1	2	2

References

Alves JABLR, Dantas ERS, Pergher SBC et al (2014) Synthesis of high value-added zeolitic materials using glass powder residue as a silica source. Mater Res 17:213–218. https://doi.org/10.1590/S1516-14392013005000191

Baerlocher C, McCusker LB (2007) Database of zeolite structures. International Zeolite Association, IZA. http://www.iza-structure.org/databases/

Bieseki L, Penha FG, Pergher SBC (2012) Zeolite A synthesis employing a Brazilian coal ash as the silicon and aluminum source and its applications in adsorption and pigment formulation. Mater Res 16(1):38–43

da Silva Filho SH, Vinaches P, Pergher SBC (2018) Zeolite synthesis in basic media using expanded perlite and its application in Rhodamine B adsorption. Mater Lett 227:258–260. https://doi.org/10.1016/j.matlet.2018.05.095

Mignoni MIL, Petkowicz DI, Fernandes Machado NRC, Pergher SBC (2008) Synthesis of mordenite using kaolin as Si and Al source. Appl Clay Sci 41(1–2):99–104

Mignoni MIL, Detoni C, Pergher SBC (2007) Study of the ZSM-5 zeolite synthesis from natural clays. Quim Nova 30(1):45–48

Mintova S (ed) (2016) Verified syntheses of zeolitic materials, XRD Patterns: Barrier N. 3rd rev edn. Published on behalf of the Synthesis Commission of the International Zeolite Association 2016

Oliveira MSM, Nascimento RM, Pergher SBC (2018) Síntese de zeólita LPM-12 (tipo EDI) utilizando resíduo do processamento do espodumênio como fonte alternativa de silício e alumínio. Perspectiva, Erechim

Petkowicz D, Rigo R, Radtke C et al (2008) Zeolite NaA from Brazilian chrysotile and rice husk. Microporous Mesoporous Mater 116(1–3):548–554

Rigo RT, Pergher SBC, Petkowicz DI, dos Santos JHZ (2009) Um novo procedimento de síntese da zeólita A empregando argilas naturais. Quim Nova 32(1):21–25

Vinaches P, Alves JABLR, Melo DMA, Pergher SBC (2016) Raw powder glass as a silica source in the synthesis of colloidal MEL zeolite. Mater Lett 178:217–220. https://doi.org/10.1016/j.matlet.2016.05.006

Vinaches P, Rebitski EP, Alves JABLR et al (2015) Unconventional silica source employment in zeolite synthesis: raw powder glass in MFI synthesis case study. Mater Lett 159:233–236. https://doi.org/10.1016/j.matlet.2015.06.120

Conclusions
Future Approaches

Zeolite materials are very interesting porous solids. These materials can be synthesized in different structures, with different porosities, areas, and chemical compositions, using commercial reagents or natural raw materials such as Si and Al sources. It is possible to synthesize zeolites using Si and Al sources from waste. These characteristics made these materials ideal for several processes, such as for catalysts, adsorption materials, and molecular sieves.

Thus, it is very important study the synthesis of this kind of material to control its properties and design materials for specific applications.

Currently, only 11 zeolites (from 245 known) are produced industrially, which is due to the high costs of synthesis. Thus, eco-friendly synthesis is a very good way to achieve different industrial zeolites.

In the near future, the goal will be to achieve more economic and eco-friendlier zeolite syntheses.

© Springer Nature Switzerland AG 2019
R. Chaves Lima et al., *Environmentally Friendly Zeolites*, Engineering Materials,
https://doi.org/10.1007/978-3-030-19970-8

Printed in the United States
By Bookmasters